Dimensions Math®
Teacher's Guide 2A

Authors and Reviewers

Cassandra Turner

Allison Coates

Jenny Kempe

Bill Jackson

Tricia Salerno

Singapore Math Inc.

Published by Singapore Math Inc.

19535 SW 129th Avenue
Tualatin, OR 97062
www.singaporemath.com

Dimensions Math® Teacher's Guide 2A
ISBN 978-1-947226-34-0

First published 2018
Reprinted 2019, 2020

Printed in China

Acknowledgments

Editing by the Singapore Math Inc. team.
Design and illustration by Cameron Wray with Carli Fronius.

Contents

Chapter		Lesson	Page

Dimensions Math® Curriculum

The **Dimensions Math®** series is a Pre-Kindergarten to Grade 5 series based on the pedagogy and methodology of math education in Singapore. The main goal of the **Dimensions Math®** series is to help students develop competence and confidence in mathematics.

The series follows the principles outlined in the Singapore Mathematics Framework below.

Pedagogical Approach and Methodology

- Through Concrete-Pictorial-Abstract development, students view the same concepts over time with increasing levels of abstraction.
- Thoughtful sequencing creates a sense of continuity. The content of each grade level builds on that of preceding grade levels. Similarly, lessons build on previous lessons within each grade.
- Group discussion of solution methods encourages expansive thinking.
- Interesting problems and activities provide varied opportunities to explore and apply skills.
- Hands-on tasks and sharing establish a culture of collaboration.
- Extra practice and extension activities encourage students to persevere through challenging problems.
- Variation in pictorial representation (number bonds, bar models, etc.) and concrete representation (straws, linking cubes, base ten blocks, discs, etc.) broaden student understanding.

Each topic is introduced, then thoughtfully developed through the use of a variety of learning experiences, problem solving, student discourse, and opportunities for mastery of skills. This combination of hands-on practice, in-depth exploration of topics, and mathematical variability in teaching methodology allows students to truly master mathematical concepts.

Singapore Mathematics Framework

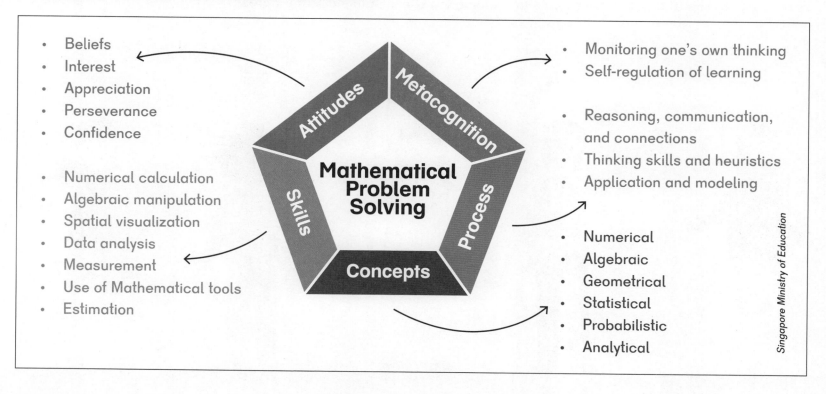

Singapore Ministry of Education

Dimensions Math® Program Materials

Textbooks

Textbooks are designed to help students build a solid foundation in mathematical thinking and efficient problem solving. Careful sequencing of topics, well-chosen problems, and simple graphics foster deep conceptual understanding and confidence. Mental math, problem solving, and correct computation are given balanced attention in all grades. As skills are mastered, students move to increasingly sophisticated concepts within and across grade levels.

Students work through the textbook lessons with the help of five friends: Emma, Alex, Sofia, Dion, and Mei. The characters appear throughout the series and help students develop metacognitive reasoning through questions, hints, and ideas.

A pencil icon ▬▬▬▶ at the end of the textbook lessons links to exercises in the workbooks.

Workbooks

Workbooks provide additional problems that range from basic to challenging. These allow students to independently review and practice the skills they have learned.

Teacher's Guides

Teacher's Guides include lesson plans, mathematical background, games, helpful suggestions, and comprehensive resources for daily lessons.

Tests

Tests contain differentiated assessments to systematically evaluate student progress.

Emma   Alex   Sofia   Dion   Mei

Online Resources

The following can be downloaded from dimensionsmath.com.

- **Blackline Masters** used for various hands-on tasks.

- **Material Lists** for each chapter and lesson, so teachers and classroom helpers can prepare ahead of time.

- **Activities** that can done with students who need more practice or a greater challenge, organized by concept, chapter, and lesson.

- **Standards Alignments** for various states.

Using the Teacher's Guide

This guide is designed to assist in planning daily lessons. It should be considered a helping hand between the curriculum and the classroom. It provides introductory notes on mathematical content, key points, and suggestions for activities. It also includes ideas for differentiation within each lesson, and answers and solutions to textbook and workbook problems.

Each chapter of the guide begins with the following.

- ## Overview

  Includes objectives and suggested number of class periods for each chapter.

- ## Notes

  Highlights key learning points, provides background on math concepts, explains the purpose of certain activities, and helps teachers understand the flow of topics throughout the year.

- ## Materials

  Lists materials, manipulatives, and Blackline Masters used in the Think and Learn sections of the guide. It also includes suggested storybooks. Many common classroom manipulatives are used throughout the curriculum. When a lesson refers to a whiteboard and markers, any writing materials can be used. Blackline Masters can be found at dimensionsmath.com.

The guide goes through the Chapter Openers, Daily Lessons, and Practices of each chapter, and cumulative reviews in the following general format.

- ## Chapter Opener

  Provides talking points for discussion to prepare students for the math concepts to be introduced.

- ## Think

  Offers structure for teachers to guide student inquiry. Provides various methods and activities to solve initial textbook problems or tasks.

- ## Learn

  Guides teachers to analyze student methods from Think to arrive at the main concepts of the lesson through discussion and study of the pictorial representations in the textbook.

- ## Do

  Expands on specific problems with strategies, additional practice, and remediation.

● Activities

Allows students to practice concepts through individual, small group, and whole group hands-on tasks and games, including suggestions for outdoor play (most of which can be modified for a gymnasium or classroom).

Level of difficulty in the games and activities are denoted by the following symbols.

- ● Foundational activities
- ▲ On-level activities
- ★ Challenge or extension activities

● Brain Works

Provides opportunities for students to extend their mathematical thinking.

Discussion is a critical component of each lesson. Teachers are encouraged to let students discuss their reasoning. As each classroom is different, this guide does not anticipate all situations. The following questions can help students articulate their thinking and increase their mastery:

- Why? How do you know?
- Can you explain that?
- Can you draw a picture of that?
- Is your answer reasonable? How do you know?
- How is this task like the one we did before? How is it different?
- What is alike and what is different about…?
- Can you solve that a different way?
- Yes! You're right! How do you know it's true?
- What did you learn before that can help you solve this problem?
- Can you summarize what your classmate shared?
- What conclusion can you draw from the data?

Each lesson is designed to take one day. If your calendar allows, you may choose to spend more than one day on certain lessons. Throughout the guide, there are notes to extend on learning activities to make them more challenging. Lesson structures and activities do not have to conform exactly to what is shown in the guide. Teachers are encouraged to exercise their discretion in using this material in a way that best suits their classes.

Textbooks are designed to last multiple years. Textbook problems with a ▇ (or a blank line for terms) are meant to invite active participation.

Dimensions Math® Scope & Sequence

Dimensions Math® Scope & Sequence

Count Up to 10 Things —
Part 2
Recognize the Numbers
6 to 10
Write the Numbers 6 and 7
Write the Numbers 8, 9,
and 10
Write the Numbers 6 to 10
Count and Write the
Numbers 1 to 10
Ordinal Positions
One More Than
Practice

Chapter 4
Shapes and Solids

Curved or Flat
Solid Shapes
Closed Shapes
Rectangles
Squares
Circles and Triangles
Where is It?
Hexagons
Sizes and Shapes
Combine Shapes
Graphs
Practice

Chapter 5
Compare Height, Length, Weight, and Capacity

Comparing Height
Comparing Length
Height and Length — Part 1
Height and Length — Part 2
Weight — Part 1

Weight — Part 2
Weight — Part 3
Capacity — Part 1
Capacity — Part 2
Practice

Chapter 6
Comparing Numbers Within 10

Same and More
More and Fewer
More and Less
Practice — Part 1
Practice — Part 2

KB

Chapter 7
Numbers to 20

Ten and Some More
Count Ten and Some More
Two Ways to Count
Numbers 16 to 20
Number Words 0 to 10
Number Words 11 to 15
Number Words 16 to 20
Number Order
1 More Than or Less Than
Practice — Part 1
Practice — Part 2

Chapter 8
Number Bonds

Putting Numbers Together
— Part 1

Putting Numbers Together
— Part 2
Parts Making a Whole
Look for a Part
Number Bonds for 2, 3, and 4
Number Bonds for 5
Number Bonds for 6
Number Bonds for 7
Number Bonds for 8
Number Bonds for 9
Number Bonds for 10
Practice — Part 1
Practice — Part 2
Practice — Part 3

Chapter 9
Addition

Introduction to Addition —
Part 1
Introduction to Addition —
Part 2
Introduction to Addition —
Part 3
Addition
Count On — Part 1
Count On — Part 2
Add Up to 3 and 4
Add Up to 5 and 6
Add Up to 7 and 8
Add Up to 9 and 10
Addition Practice
Practice

Chapter 10
Subtraction

Take Away to Subtract —
Part 1

Dimensions Math® Scope & Sequence

Dimensions Math® Scope & Sequence

Dividing by 5 and 10
Practice C
Word Problems
Review 2

2B

Chapter 8
Mental Calculation

Adding Ones Mentally
Adding Tens Mentally
Making 100
Adding 97, 98, or 99
Practice A
Subtracting Ones Mentally
Subtracting Tens Mentally
Subtracting 97, 98, or 99
Practice B
Practice C

Chapter 9
Multiplication and Division of 3 and 4

The Multiplication Table of 3
Multiplication Facts of 3
Dividing by 3
Practice A
The Multiplication Table of 4
Multiplication Facts of 4
Dividing by 4
Practice B
Practice C

Chapter 10
Money

Making $1
Dollars and Cents
Making Change
Comparing Money
Practice A
Adding Money
Subtracting Money
Practice B

Chapter 11
Fractions

Halves and Fourths
Writing Unit Fractions
Writing Fractions
Fractions that Make 1 Whole
Comparing and Ordering
    Fractions
Practice
Review 3

Chapter 12
Time

Telling Time
Time Intervals
A.M. and P.M.
Practice

Chapter 13
Capacity

Comparing Capacity
Units of Capacity
Practice

Chapter 14
Graphs

Picture Graphs
Bar Graphs
Practice

Chapter 15
Shapes

Straight and Curved Sides
Polygons
Semicircles and Quarter-
    circles
Patterns
Solid Shapes
Practice
Review 4
Review 5

3A

Chapter 1
Numbers to 10,000

Numbers to 10,000
Place Value — Part 1
Place Value — Part 2
Comparing Numbers
The Number Line
Practice A
Number Patterns
Rounding to the Nearest
    Thousand
Rounding to the Nearest
    Hundred
Rounding to the Nearest Ten
Practice B

Dimensions Math® Scope & Sequence

Dimensions Math® Scope & Sequence

Dimensions Math® Scope & Sequence

Conversion of Measures
Mental Calculation
Practice B

Chapter 10
The Four Operations of Decimals

Adding Decimals to Thousandths
Subtracting Decimals
Multiplying by 0.1 or 0.01
Multiplying by a Decimal
Practice A
Dividing by a Whole Number — Part 1
Dividing by a Whole Number — Part 2
Dividing a Whole Number by 0.1 and 0.01
Dividing a Whole Number by a Decimal
Practice B

Chapter 11
Geometry

Measuring Angles
Angles and Lines
Classifying Triangles
The Sum of the Angles in a Triangle
The Exterior Angle of a Triangle
Classifying Quadrilaterals
Angles of Quadrilaterals — Part 1
Angles of Quadrilaterals — Part 2

Drawing Triangles and Quadrilaterals
Practice

Chapter 12
Data Analysis and Graphs

Average — Part 1
Average — Part 2
Line Plots
Coordinate Graphs
Straight Line Graphs
Practice
Review 3

Chapter 13
Ratio

Finding the Ratio
Equivalent Ratios
Finding a Quantity
Comparing Three Quantities
Word Problems
Practice

Chapter 14
Rate

Finding the Rate
Rate Problems — Part 1
Rate Problems — Part 2
Word Problems
Practice

Chapter 15
Percentage

Meaning of Percentage
Expressing Percentages as Fractions

Percentages and Decimals
Expressing Fractions as Percentages
Practice A
Percentage of a Quantity
Word Problems
Practice B
Review 4
Review 5

Suggested number of class periods: 8–9

	Lesson	Page	Resources		Objectives
	Chapter Opener	p. 5	TB:	p. 1	Investigate numbers to 1,000.
1	Tens and Ones	p. 6	TB: WB:	p. 2 p. 1	Decompose a two-digit number to tens and ones.
2	Counting by Tens or Ones	p. 9	TB: WB:	p. 7 p. 3	Find the number that is 1, 2, 3, 10, 20, or 30 more or less than a number within 100.
3	Comparing Tens and Ones	p. 11	TB: WB:	p. 10 p. 5	Compare two-digit numbers and represent their relationship using >, <, and = signs. Order up to 4 two-digit numbers within 100.
4	Hundreds, Tens, and Ones	p. 14	TB: WB:	p. 14 p. 7	Understand how three-digit numbers are composed according to place value. Count quantities up to 1,000 by counting by hundreds, tens, and ones.
5	Place Value	p. 17	TB: WB:	p. 18 p. 11	Represent three-digit numbers on a place-value chart. Write three-digit numbers in expanded form.
6	Comparing Hundreds, Tens, and Ones	p. 21	TB: WB:	p. 23 p. 15	Compare three-digit numbers by comparing hundreds, tens, and ones.
7	Counting by Hundreds, Tens, or Ones	p. 24	TB: WB:	p. 29 p. 19	Find the number that is 1, 10, or 100 more or less than a number within 1,000.
8	Practice	p. 27	TB: WB:	p. 33 p. 23	Practice writing and comparing numbers within 1,000.
	Workbook Solutions	p. 29			

In **Dimensions Math® 1**, students learned to:

- Relate two-digit numbers to place value.
- Use base ten blocks and linking cubes to show a two-digit number by place value and use place-value charts to form numbers.
- Compare 2 two-digit numbers using the terms "greater than" and "less than."

In this chapter, students build upon their knowledge of two-digit numbers and extend concepts and skills to three-digit numbers. In order to gain a solid understanding of place value, students should have sufficient hands-on experience with manipulatives and see many different representations of place value.

While working with place value, students will use several tools, including objects that can be bundled into tens, base ten blocks, number cards, and place-value charts. Base ten blocks are slightly more abstract than the linking cubes students used in grade 1 because the ten rods do not come apart into individual ones in most sets.

Students who struggle to understand that a ten rod still has 10 units should use base ten blocks. Base ten blocks allow students to count each unit and gain a better understanding of magnitude of numbers.

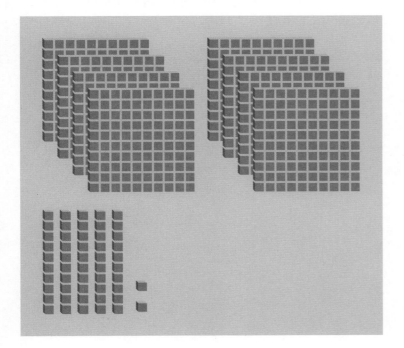

Place-Value Discs

This chapter introduces the use of place-value discs, which will be used throughout this series for whole numbers and decimal numbers. These are more abstract than base ten blocks as they are not proportional in size. Students must take as a fact that a 10-disc represents the value of 10 ones and a 100-disc represents the value of 10 tens. Students who are familiar with coins and understand that a dime has the same value as ten pennies should have little difficulty in grasping that a 10-disc has the same value as ten 1-discs.

Place-value discs better mirror our base ten number system, which is based on place value. The position, or place of the digit, with respect to other digits, determines its value. Each place represents a value ten times the place to its left.

Numbers greater than hundreds are easier to represent with place-value discs than base ten blocks. Students familiar with place-value discs can easily extend the concept to decimal numbers in **Dimensions Math® 4B**.

Finally, place-value discs can also be easier to work with on a student desk. Once students transition to place-value discs, base ten blocks will still be used periodically when introducing new concepts.

In 2A, students will use a place-value chart divided into hundreds, tens, and ones. This expands on the tens and ones charts used in **Dimensions Math® 1B**. In 3A, students will add a thousands column to the chart. Students write numbers in the correct column according to place.

Using place-value manipulatives **can** be confusing at first. In order to avoid confusion, a place-value chart uses headers. This way, students won't see a disc with 10 on it in a tens column and think it is worth 10 tens (instead of 1 ten).

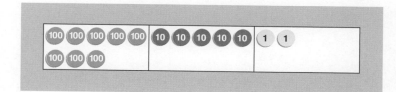

Hundreds	Tens	Ones
8	5	2

If discs are used, there are no headers indicating which column is hundreds, tens, or ones.

Students should have a way of organizing their discs by place and differentiating them from other discs on their desks. They can do this by simply creating 3 distinct areas on their desks using a paper mat divided into columns.

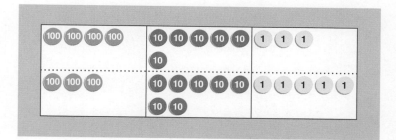

A student or pair of students will need 20 each of tens and ones discs and 10 hundreds discs.

Place-Value Cards

These manipulatives are provided as a Blackline Master. Students should have their own sets up to the hundreds.

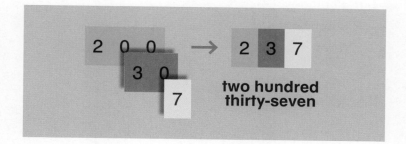

two hundred thirty-seven

Place-value cards help students understand composing and decomposing numbers into their corresponding place values.

Addition and Subtraction

In this chapter, students will trade 10 ones for 1 ten and 10 tens for 1 hundred to facilitate their understanding of 10 more, 10 less, 100 more, and 100 less. Students will learn regrouping for addition and subtraction in **Chapter 3: Addition and Subtraction — Part 2**.

In **Dimensions Math® 1**, students used the phrases "greater than," "less than," or "equal to" to compare values. In this chapter, students will now learn the mathematical symbols:

Greater than   >
Less than   <

To avoid confusion, this series does not use imagery, such as an open crocodile's mouth, to stand in for mathematical symbols.

Materials

- Base ten blocks
- Place-value discs
- Place-value organizers
- Play money
- Rubber bands
- Straws

Blackline Masters

- Count by Ones or Tens Game Cards
- Number Cards 1 to 100
- Place-value Cards
- Two-Digit Find Your Match Cards
- Three-Digit Find Your Match Cards

Storybooks

- *Place Value* by David A. Adler
- *Earth Day — Hooray* by Stuart J. Murphy
- *Sir Cumference and All the King's Tens* by Cindy Neuschwander

Activities

Games and activities included in this chapter are designed to provide practice and extensions of place-value concepts. They can be used after students complete the **Do** questions, or anytime review and practice are needed.

Chapter Opener

Objective

- Investigate numbers to 1,000.

Lesson Materials

- Straws or craft sticks, between 100–200 per group
- Rubber bands

Provide groups of students with between 100 and 200 straws. Ask them to guess how many straws they have and discuss how to count them.

Lead students to see that it is easier to count the straws if they are grouped in tens. Have them bundle the straws in tens and ask what should be done with the remaining straws.

Ask students if the bundled straws can be further bundled to make it even easier to count.

Lead students to bundle 10 tens into 1 hundred. Then have each group of students count their straws. For example, if they have 123 straws they can say, "100, 110, 120, 121, 122, 123."

Keep the bundled straws for use in **Lesson 1: Tens and Ones**.

Lesson 1 Tens and Ones

Objective

- Decompose a two-digit number to tens and ones.

Lesson Materials

- Place-value Cards (BLM)

Think

Draw a place-value chart on the board. Have students count the number of straws in **Think** and share their answers on whiteboards.

Discuss that the written numeral shows the number of tens and the number of ones from left to right. Ask them if they can also show this with a number bond where one part is the tens and one part is the ones.

Provide students with Place-value Cards (BLM) and have them show the number of straws with tens and ones. Have students show tens and ones for different two-digit numbers.

Learn

Have students discuss the different representations of the number 78. Have them show with their Place-value Cards (BLM) how two-digit numbers are composed of tens and ones.

Continue to provide different two-digit numbers and have students represent them or say them in different ways, similar to the **Learn** examples.

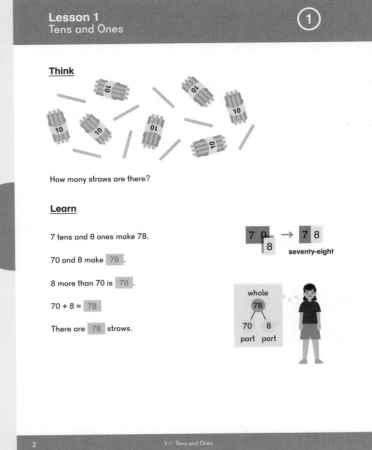

Lesson 1
Tens and Ones

Think

How many straws are there?

Learn

7 tens and 8 ones make 78.

70 and 8 make 78 .

8 more than 70 is 78 .

70 + 8 = 78

There are 78 straws.

7 0 → 7 8
8

seventy-eight

whole
78
70   8
part part

2   1-1 Tens and Ones

Do

These questions should be review. If students are struggling, consider using lessons and games from Chapter 16 in **Dimensions Math® 1B**.

1 These problems show the various expressions for adding tens and ones. Remind students that whichever expression is used, the answer is the same, and the pattern to first show the tens and then the ones starting from the left remains the same.

1 (d) Alex is reviewing counting by tens.

2 Students should use the Place-value Cards (BLM).

Activities

▲ Greatest or Least?

Materials: Number Cards (BLM) 1 to 100

This game works well with up to 6 players.

Shuffle and place about half of the cards facedown in front of players.

Each player draws a card. The players with the greatest and least numbers on their cards keep theirs. The other players return their cards to the pile.

The round continues until there are no cards left. Players can play another round with the remaining cards, or reshuffle all cards and lay another group of cards facedown.

The player with the most cards at the end of two rounds wins the game.

This activity reviews **Dimensions Math® 1B Chapter 16, Lesson 4** by comparing tens and ones.

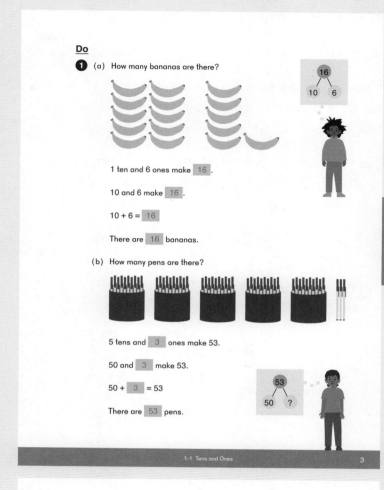

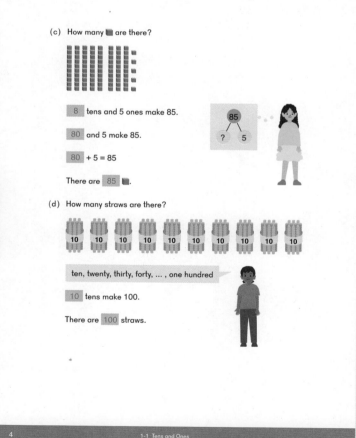

▲ Find Your Match — Two-Digit

Materials: Two-Digit Find Your Match Cards (BLM)

This can be a whole class activity that uses number words, numbers in expanded form, number bonds, and place-value chart representations.

Pass out one card to each student and have them find cards with numbers or representations that match their card.

Students can also line up in order from least to greatest, with students holding cards with the same number standing in front of or behind their number match.

| thirty-five | 35 | (number bond: 35 = 30 and 5) |

Exercise 1 • page 1

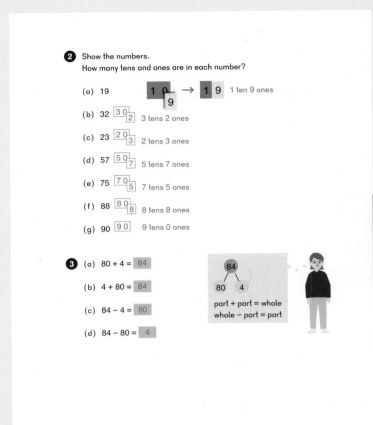

2 Show the numbers.
How many tens and ones are in each number?

(a) 19   1 0 → 1 9   1 ten 9 ones
     9

(b) 32   3 0 2   3 tens 2 ones

(c) 23   2 0 3   2 tens 3 ones

(d) 57   5 0 7   5 tens 7 ones

(e) 75   7 0 5   7 tens 5 ones

(f) 88   8 0 8   8 tens 8 ones

(g) 90   9 0   9 tens 0 ones

3 (a) 80 + 4 = 84

(b) 4 + 80 = 84

(c) 84 − 4 = 80

(d) 84 − 80 = 4

84
80   4

part + part = whole
whole − part = part

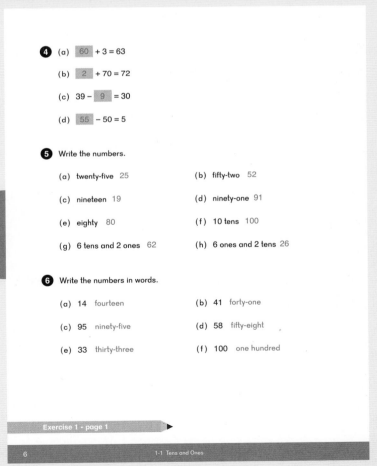

4 (a) 60 + 3 = 63

(b) 2 + 70 = 72

(c) 39 − 9 = 30

(d) 55 − 50 = 5

5 Write the numbers.

(a) twenty-five 25   (b) fifty-two 52

(c) nineteen 19   (d) ninety-one 91

(e) eighty 80   (f) 10 tens 100

(g) 6 tens and 2 ones 62   (h) 6 ones and 2 tens 26

6 Write the numbers in words.

(a) 14 fourteen   (b) 41 forty-one

(c) 95 ninety-five   (d) 58 fifty-eight

(e) 33 thirty-three   (f) 100 one hundred

Exercise 1 • page 1

5

6

Lesson 2 Counting by Tens or Ones

Objective

- Find the number that is 1, 2, 3, 10, 20, or 30 more or less than a number within 100.

Lesson Materials

- Base ten blocks

Think

Provide students with base ten blocks and pose the problem in **Think**. Students can then write the numbers on whiteboards.

Discuss what happened to the written number for 3 more than 48 (51).

Ask students what the ten digits are that we use to represent written numbers. Tell them that these same digits are used to represent all numbers. Ask them how they would represent 3 more than 48 as a number.

Discuss which digit changes when finding 30 more than 48 and 20 less than 48.

Learn

Have students discuss how the five friends are adding and subtracting.

Students should work the problems with base ten blocks. When adding 3 to 48, they will need to count 49, 50, 51, trade 10 ones for 1 ten. 51 is 5 tens and 1 one.

In Dion and Alex's problems, students should note that when counting on and back by tens, they can simply add or take away ten rods.

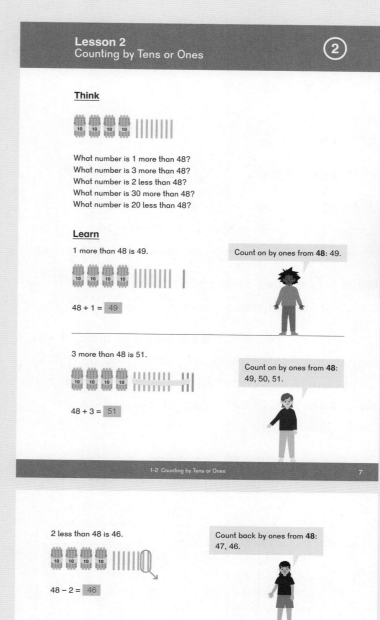

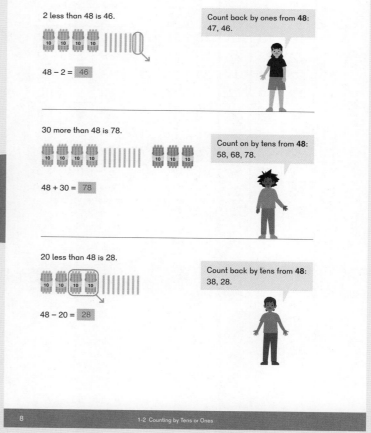

Do

Discuss with students how the digits change as ones and tens are added or subtracted.

Activity

▲ **Three in a Row**

Materials: Number Cards (BLM) 0 to 9 or playing cards (face cards removed), Count by Ones or Tens Game Cards (BLM), hundred chart, game markers

Players take turns drawing two Number Cards (BLM) and a Count by Ones or Tens Game Card (BLM).

On each turn, players make a two-digit number with the playing cards. They then perform the operation on the Count by Ones or Tens Game Card (BLM) and place their marker on the corresponding number on the hundred chart.

The first player to mark 3 numbers in a row, (horizontally, vertically, or diagonally) wins.

1	2	3	4	5	6	7	8	9	10
11	12	13	14	15	16	17	18	19	20
21	22	23	24	25	26	27	28	29	30
31	32	33	34	35	36	37	38	39	40
41	42	43	44	45	46	47	48	49	50
51	52	53	54	55	56	57	58	59	60
61	62	63	64	65	66	67	68	69	70
71	72	73	74	75	76	77	78	79	80
81	82	83	84	85	86	87	88	89	90
91	92	93	94	95	96	97	98	99	100

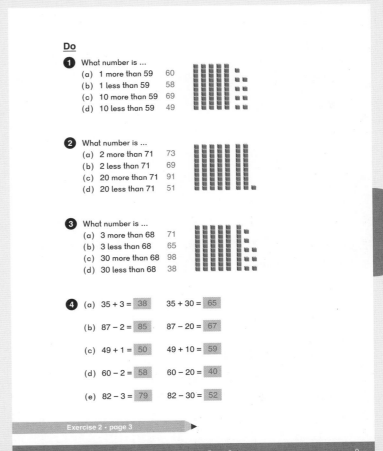

Exercise 2 • page 3

9

Teacher's Guide 2A Chapter 1 © 2017 Singapore Math Inc.

Objectives

- Compare two-digit numbers and represent their relationship using >, <, and = signs.
- Order up to 4 two-digit numbers within 100.

Lesson Materials

- Base ten blocks
- Number Cards (BLM) 1 to 100

Think

Provide students with base ten blocks and pose the problem in **Think**.

Have students explain why one number is greater than another. Most students may answer simply that when you count, 28 comes first. Ask them if there is a quick way of comparing the numbers by simply looking at the digits in the numbers, and knowing the order of numbers 0 to 9.

Prompt students to use the terms "greater than" and "less than."

Learn

Write on the board, "Thirty-two is greater than twenty-eight."

Explain that it takes a longer time to write words than to write symbols. Just as there are symbols for number words, there is a symbol for the words "is greater than."

Write 32 > 28 on the board.

Change the order in which the numbers are shown by writing "28" and "32" on the board and repeat the process to explain "less than."

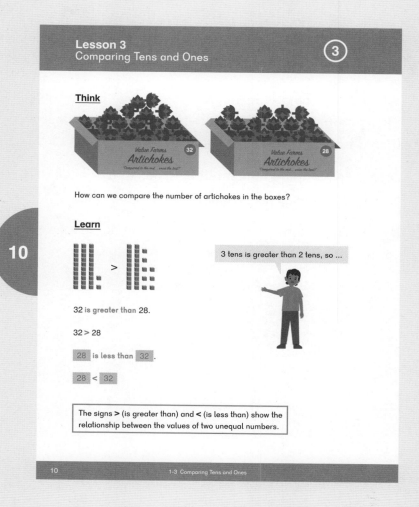

Note: If students have trouble remembering which way the symbol should "point," have them make two dots next to the greater number and one dot next to the lesser number. When students connect the dots, they create the correct sign.

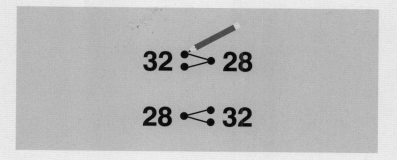

Be sure to always read the sentence with the correct language:

52 < 85   Fifty-two is **less than** eighty-five.
85 > 52   Eighty-five is **greater than** fifty-two.

Do

1 Discuss Emma and Dion's strategies.

Students are guided to look at the greatest place value to compare amounts. Emma thinks that 4 tens is less than 6 tens, so she knows that 49 is less than 61.

Dion notes that the tens are the same, so he needs to look to the ones to see which number is greater.

2 Mei reminds students that the equal sign means the values, or amounts, are equal.

Students should read this as, "75 is equal to 75."

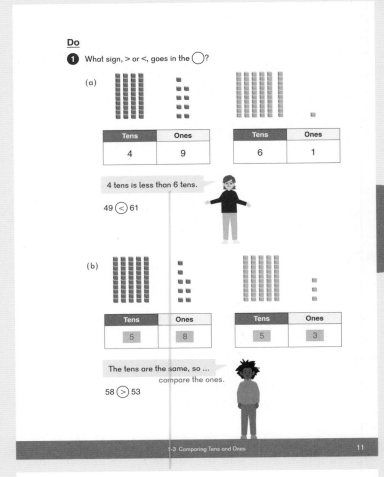

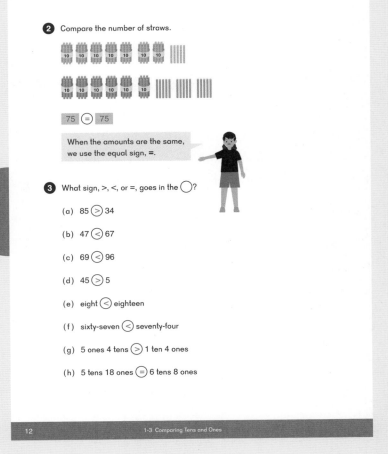

11

12

④ Cards can be pulled from Number Cards (BLM) 1 to 100 to use this question in a center or as a whole class activity. Have students come to the front of the class and face the class, each holding a card. The class tells them how to rearrange themselves in order from least to greatest.

⑤ Students may need to re-write some of these problems to solve.

Activity

▲ More or Fewer Face-off

Materials: Playing cards with face cards removed (include the Jokers as 0), linking cubes or counters

Players each flip over two cards at the same time. The first card is the tens and the second card is the ones.

The greatest number (or least) wins.

● Students can use linking cubes or counters to see whose number is greater.

★ Have the winner say how many greater or less their number is than the other player's number.

Exercise 3 · page 5

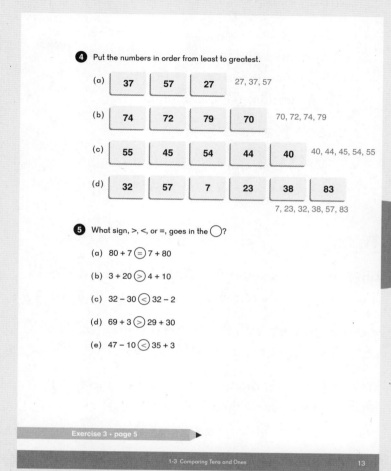

④ Put the numbers in order from least to greatest.

(a) 37   57   27   27, 37, 57

(b) 74   72   79   70   70, 72, 74, 79

(c) 55   45   54   44   40   40, 44, 45, 54, 55

(d) 32   57   7   23   38   83
     7, 23, 32, 38, 57, 83

⑤ What sign, >, <, or =, goes in the ◯?

(a) 80 + 7 (=) 7 + 80

(b) 3 + 20 (>) 4 + 10

(c) 32 – 30 (<) 32 – 2

(d) 69 + 3 (>) 29 + 30

(e) 47 – 10 (<) 35 + 3

Exercise 3 · page 5

1-3 Comparing Tens and Ones   13

13

Lesson 4 Hundreds, Tens, and Ones

Objectives

- Understand how three-digit numbers are composed according to place value.
- Count quantities up to 1,000 by counting by hundreds, tens, and ones.

Lesson Materials

- Place-value Cards (BLM) to the hundreds

Think

Have students discuss the **Think** question and ask them how to write a number that is more than 10 tens.

Provide students with Place-value Cards (BLM) and have them show the number of straws.

Write other three-digit numbers on the board and have students show them with the cards.

Learn

Have students discuss the different representations of the number 125.

Continue to provide different three-digit numbers and have students:

- Show the numbers with Place-value Cards (BLM).
- Represent the numbers in hundreds, tens, and ones.
- Say the numbers in different ways, similar to the **Learn** examples.

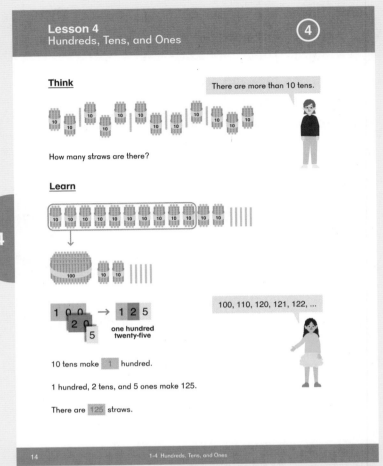

Do

Guide students to count from the greatest place value the way Sofia and Alex demonstrate.

Students can practice writing the number words.

Note: Explain that numbers in the hundreds do not have special names like tens numbers do. There is the word "forty" for 4 tens, but only "four hundred."

1 (c) Note that in the number 406 there are no tens, and students should use the 400 strip, not one that reads 00.

1 (d) When writing the number for 10 hundreds, we need a new place. Thousands are introduced here.

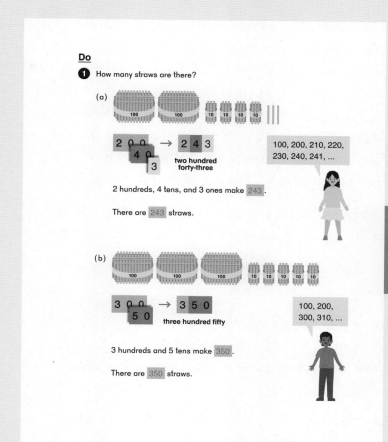

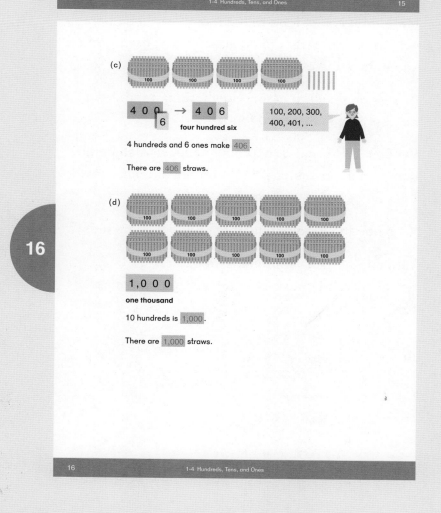

2 The number is always written in the order of hundreds, tens, and ones, regardless of any physical or pictorial representation, or the order in which the number of ones, tens, or hundreds is said or written as individual groups. Whether the ones, tens, or hundreds are first in the representation of the straws, the number of straws is still the same. Thus, 9 ones + 6 tens + 3 hundreds is 369 (not 963).

4 These problems can be challenging. Have students show the numbers with Place-value Cards (BLM).

4 These problems require students to think about place-value quantities.

After **5**, provide additional examples similar to the ones in **4** and **5** until students are comfortable composing three-digit numbers by place value.

Activities

▲ **My Name Is ...**

Materials: Name tags with decomposed numbers written on them, recording sheet listing names of all students in the class

Give each player a name tag showing a three-digit number in decomposed form.

Version 1: Provide a recording sheet with students' names and have students record the number next to their classmates' names.

Who has the greatest number? Who has the least?

Version 2: Ensure that there are 2 stickers that represent the same amount. Have students find their match.

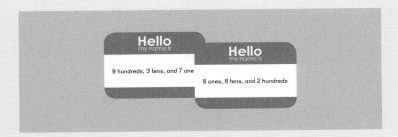

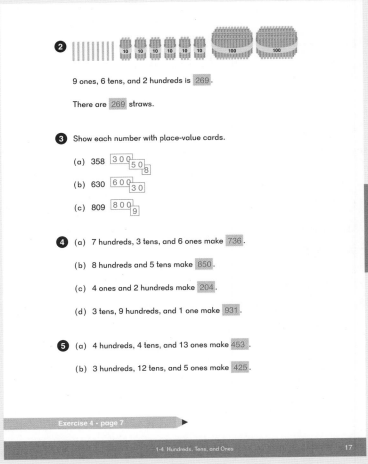

2 9 ones, 6 tens, and 2 hundreds is 269.

There are 269 straws.

3 Show each number with place-value cards.

(a) 358 300 50 8

(b) 630 600 30

(c) 809 800 9

4 (a) 7 hundreds, 3 tens, and 6 ones make 736.

(b) 8 hundreds and 5 tens make 850.

(c) 4 ones and 2 hundreds make 204.

(d) 3 tens, 9 hundreds, and 1 one make 931.

5 (a) 4 hundreds, 4 tens, and 13 ones make 453.

(b) 3 hundreds, 12 tens, and 5 ones make 425.

Exercise 4 · page 7

1-4 Hundreds, Tens, and Ones   17

▲ **Place-value Hop**

Materials: Chalk or painter's tape

Using chalk on the sidewalk, or painter's tape indoors, make a 3-row place-value board.

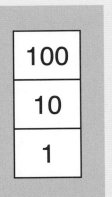

The Hopper picks a three-digit number and hops their number. Their partner needs to figure out the number.

For the number 605, the Hopper hops:

- Six times on the 100
- Over the 10
- Five times on the 1

Exercise 4 · page 7

Lesson 5 Place Value

Objectives

- Represent three-digit numbers on a place-value chart.
- Write three-digit numbers in expanded form.

Lesson Materials

- Place-value discs
- Place-value Cards (BLM)
- Base ten blocks

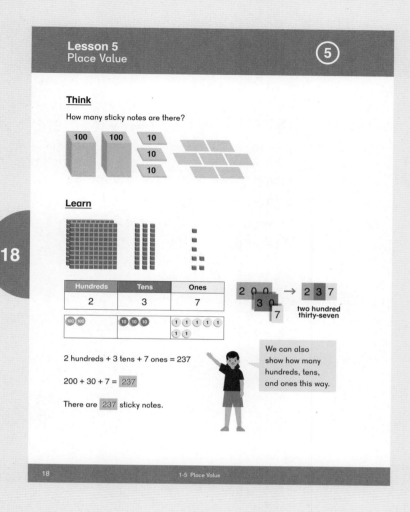

Think

Pose the problem in **Think** and ask students what strategies they know that they could use to figure out how many sticky notes there are in all.

Have students make the number with base ten blocks or Place-value Cards (BLM) and discuss the different strategies.

Hand out the place-value discs and introduce them to the students. Explain what each disc, ones, tens, and hundreds, stands for and have students make the number 237 alongside the base ten blocks.

Provide students with other numbers to represent using the place-value discs.

Learn

Have students discuss the different representations of the number 237.

Continue to provide different three-digit numbers and have students:

- Show them with Place-value Cards (BLM).
- Represent them as a sum of hundreds, tens, and ones.
- Say them in different ways, similar to the **Learn** examples.

Do

2 Students who struggle may need the discs or base ten blocks.

Most students should be working with just the pages in the textbook.

3 There are no ones in these problems. Students should note that there is no need to write 200 + 50 + 0, however we do represent 0 ones in the written number 250.

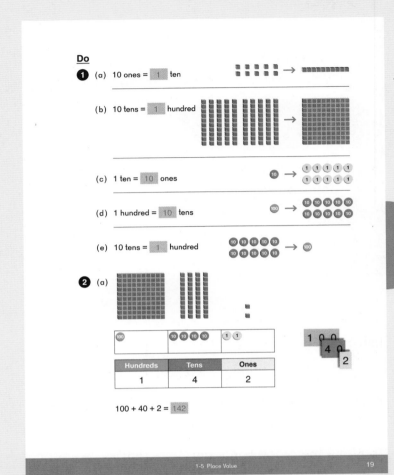

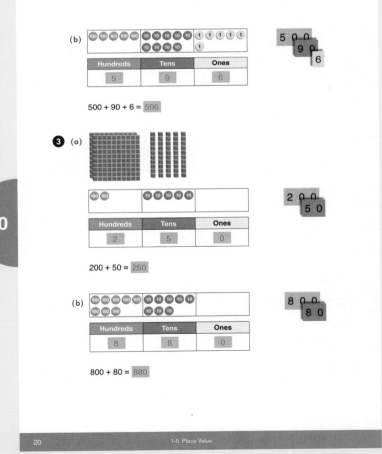

4 There are no tens in this problem. Students should note that there is no need to write 100 + 00 + 9, however, we do represent 0 tens in the written number 109.

Ask students what would happen if we didn't represent 0 tens in the written number.

5 Discuss what Alex is saying about the number 96. Generally numbers are not written with the 0 in front, i.e., 096, but sometimes numbers starting with 0 are used to name items such as dialing codes: 011, zip codes: 02134, or product names: Headphones Style No. 045.

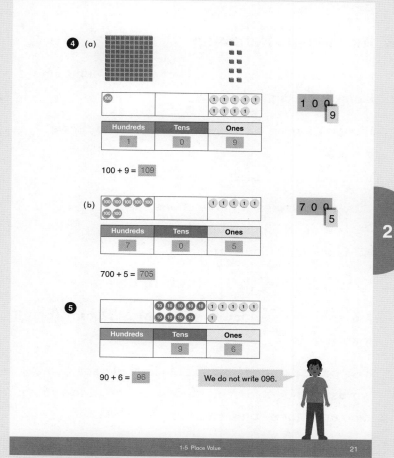

Activities

▲ **Place-value Hangman**

Have students play the traditional Hangman game using three-digit numbers.

Player 1 makes a three-digit number and draws 3 lines:

_____ _____ _____

Player 2 guesses with questions like:
- Is there a 3 in the tens place?
- Is the tens place greater than 5?

▲ **Find Your Match — Three-Digit**

Materials: Three-digit Find Your Match Cards (BLM), index cards

This can be a whole class activity that uses number words, numbers in expanded form, and place-value disc and place-value chart representations.

Pass out cards to students and have students find cards that match their own card without speaking.

★ Create additional cards with index cards.

▲ **Match Me**

Materials: Place-value Cards (BLM), Place-value discs

Students can play as partners, small groups, or the game can be played with the whole class.

One partner creates a number with the Place-value Cards (BLM). Their partner creates the same number with the place-value discs. If a third person is playing, they can write the number in numerals or expanded form.

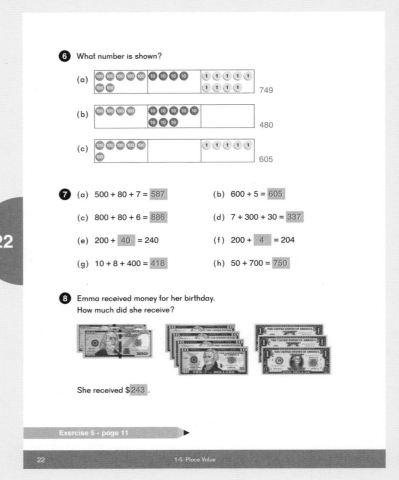

Exercise 5 · page 11

Lesson 6 Comparing Hundreds, Tens, and Ones

Objective

- Compare three-digit numbers by comparing hundreds, tens, and ones.

Lesson Materials

- Place-value discs
- Play money
- Place-value organizer

Think

Provide students with place-value discs or play money. Pose the **Think** problem.

Ask what strategies they know that they could use to figure out which game console costs more.

Discuss the different strategies.

Learn

Discuss whether students used the same strategy as Sofia did to solve the problem or a different one.

Guide students to look at the greatest place value to compare amounts. Sofia is thinking that 1 hundred is less than 2 hundreds, so $142 is less than $213.

Have students compare their solutions for **Think** with the one shown in the textbook.

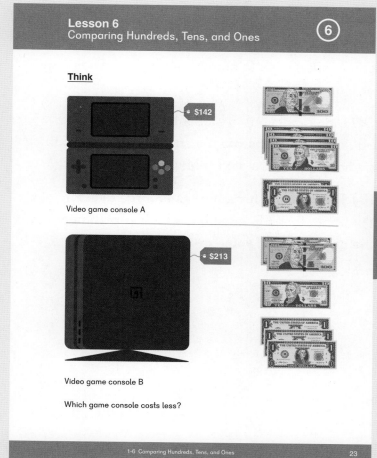

Lesson 6
Comparing Hundreds, Tens, and Ones ⑥

Think

Video game console A — $142

Video game console B — $213

Which game console costs less?

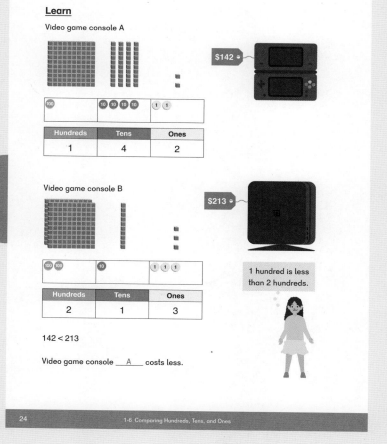

Learn

Video game console A

$142

Hundreds	Tens	Ones
1	4	2

Video game console B

$213

1 hundred is less than 2 hundreds.

Hundreds	Tens	Ones
2	1	3

142 < 213

Video game console ___A___ costs less.

Do

1 (a) Emma thinks that since the hundreds are the same, she needs to compare the tens to see if there are more tens in 251 or 238.

1 (b) Dion points out to students that both the hundreds and tens are the same, so he needs to compare the ones digit. 852 is less than 857.

1 (c) The problem uses the same three digits in different places to ensure students are comparing the correct place values.

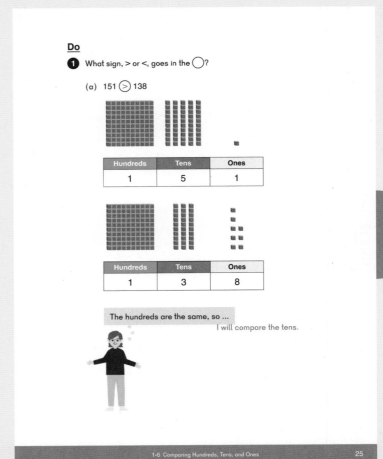

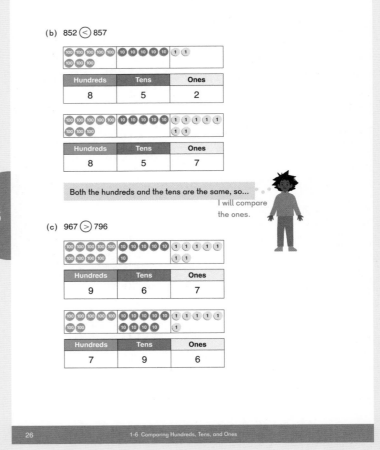

1 (d) Students should see that the number 95 has no hundreds digit, and is easy to compare to 403. 0 hundreds < 4 hundreds, so 95 is less than 403.

3 (e) Remind students that 038 is not a three-digit number.

Activities

▲ Place-value Game

Materials: Number Cards (BLM) 0 to 9 or playing cards (face cards removed)

On a whiteboard, have students draw 3 lines to represent hundreds, tens, and ones.

___ ___ ___

Students take turns drawing a card and placing it in one of the three places on their own whiteboard. After each player has created their three-digit number, the greatest number is the winner.

Shuffle the cards and play again.

The goal could also be for the lowest three-digit number.

▲ More or Fewer Face-off

Materials: Playing cards

This is similar to the game in **Lesson 3: Comparing Tens or Ones**. In this version, students make three-digit numbers. The face cards are included and are all worth zero.

Players each flip over three cards at the same time. The first card turned over is the hundreds, the second card is the tens, and the third card is the ones.

The greatest number (or least) wins.

Exercise 6 • page 15

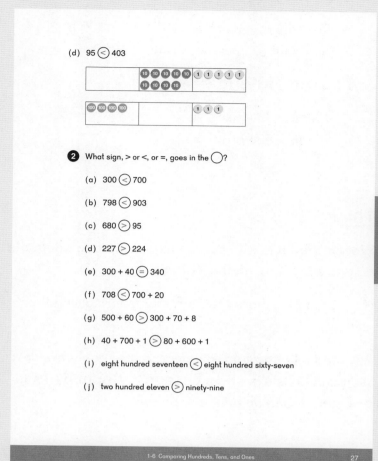

(d) 95 $<$ 403

2 What sign, > or <, or =, goes in the ◯?

(a) 300 $<$ 700

(b) 798 $<$ 903

(c) 680 $>$ 95

(d) 227 $>$ 224

(e) 300 + 40 $=$ 340

(f) 708 $<$ 700 + 20

(g) 500 + 60 $>$ 300 + 70 + 8

(h) 40 + 700 + 1 $>$ 80 + 600 + 1

(i) eight hundred seventeen $<$ eight hundred sixty-seven

(j) two hundred eleven $>$ ninety-nine

3 (a) Which number is less, 892 or 798?   798

(b) Which number is greater, 98 or 401?   401

(c) Which number is the greatest, 670, 730, or 701?   730

(d) What is the greatest number that can be made using 4, 5, and 7?   754

(e) What is the least 3-digit number that can be made using 0, 8, and 3?   308

4 Arrange the numbers in order from least to greatest.

(a) | 237 | 302 | 240 | 237, 240, 302

(b) | 74 | 702 | 715 | 517 | 74, 517, 702, 715

(c) | 207 | 72 | 720 | 270 | 27 | 27, 72, 207, 270, 720

(d) | 963 | 369 | 936 | 396 | 693 | 699 |
369, 396, 693, 699, 936, 963

Lesson 7 Counting by Hundreds, Tens, or Ones

Objective

- Find the number that is 1, 10, or 100 more or less than a number within 1,000.

Lesson Materials

- Place-value discs
- Place-value organizer

Think

Provide students with place-value discs and adequate time to work through the **Think** problem.

Have students share their strategies for solving the sticker problem.

Ask students how this is similar to the problem with Dion and his straws from **Lesson 2: Count by Tens or Ones**, found on textbook page 7.

Learn

Have students show the number of stickers that Sofia has with discs. Have them add one 100-disc to find the amount of stickers Alex has.

Have students use discs to again show the number of stickers that Sofia has, and then remove two 100-discs to find the amount of stickers that Mei has.

Write additional equations on the board where 1, 2, 3, 10, 20, 30, 100, 200, or 300 are being added or subtracted from a three-digit number and have students find the answer using place-value discs. In particular, have them do some problems where the digits in two places change, e.g., 349 + 2.

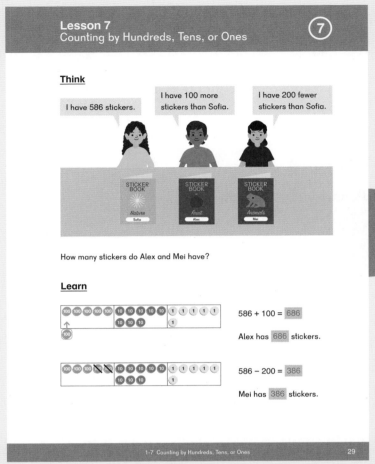

Do

Students who struggle may continue to use the discs. Most students, however, should be working with just the pages in the textbook.

Prior to working ❷, take some time to play the **Race to 100** game shown below to practice an informal introduction to regrouping.

▲ Race to 100

Materials: Place-value discs, place-value organizer, 6 or 10-sided die

Players take turns rolling the die and placing that many place-value discs on their organizers.

When the number of discs in a column is 10 (or more), students trade them for a 10-disc.

Play continues until one person collects enough discs to get to 100.

★ Extend **Race to 100** with a problem where the digits change in all three places, such as 299 + 2.

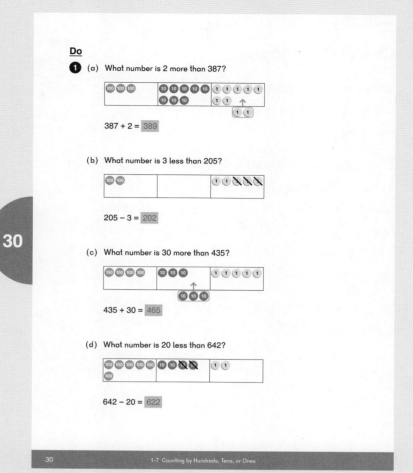

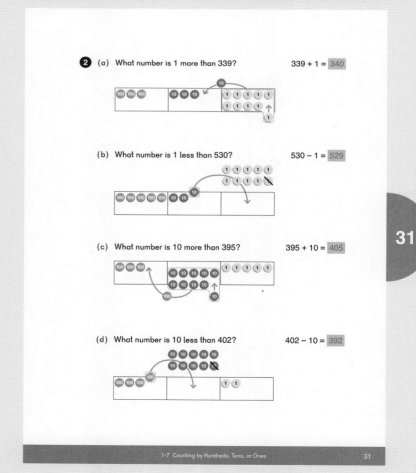

Activity

▲ **Choral Counting**

Using your thumb to point up or down, have students chorally count on and back by ones, tens, or hundreds.

Example: "Let's count by tens starting at 488, first number?" Class: "488." Point thumb up (class responds, "498"), then point up again (class responds, "508"). Point down (class responds, "498"), etc.

Exercise 7 • page 19 ▶

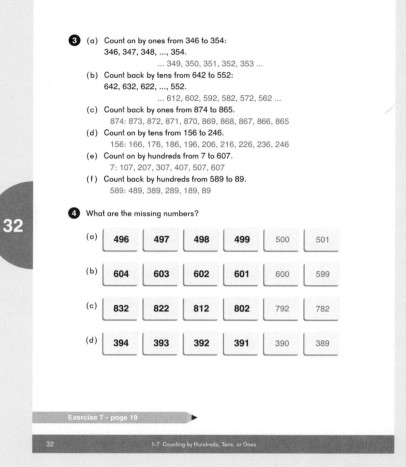

3 (a) Count on by ones from 346 to 354:
346, 347, 348, ..., 354.
... 349, 350, 351, 352, 353 ...

(b) Count back by tens from 642 to 552:
642, 632, 622, ..., 552.
... 612, 602, 592, 582, 572, 562 ...

(c) Count back by ones from 874 to 865.
874: 873, 872, 871, 870, 869, 868, 867, 866, 865

(d) Count on by tens from 156 to 246.
156: 166, 176, 186, 196, 206, 216, 226, 236, 246

(e) Count on by hundreds from 7 to 607.
7: 107, 207, 307, 407, 507, 607

(f) Count back by hundreds from 589 to 89.
589: 489, 389, 289, 189, 89

4 What are the missing numbers?

(a) | 496 | 497 | 498 | 499 | 500 | 501 |

(b) | 604 | 603 | 602 | 601 | 600 | 599 |

(c) | 832 | 822 | 812 | 802 | 792 | 782 |

(d) | 394 | 393 | 392 | 391 | 390 | 389 |

32

Exercise 7 • page 19 ▶

32   1-7 Counting by Hundreds, Tens, or Ones

Lesson 8 Practice

Objective

- Practice writing and comparing numbers within 1,000.

Practice

After students complete the **Practice** in the textbook, have them continue working with place value by playing games from the chapter.

A solid understanding of three-digit numbers is necessary before students move on to **Chapter 3: Addition and Subtraction — Part 2.**

Activities

▲ What's My Rule?

Materials: Place-value discs

Students take turns being the Ruler and the Solver.

The Ruler writes a three-digit number and a rule. The Solver writes the next five numbers that follow the rule.

For example, if the Ruler writes "377" and "+ 20," the Solver needs to write, "397, 417, 437, 457, 477."

Students can use place-value discs to check if there is disagreement.

★ Challenge students with a two-step rule. For example, the Ruler may write "387" and "+10, −1." The Solver needs to write, "396, 405, 414, 423, and 432."

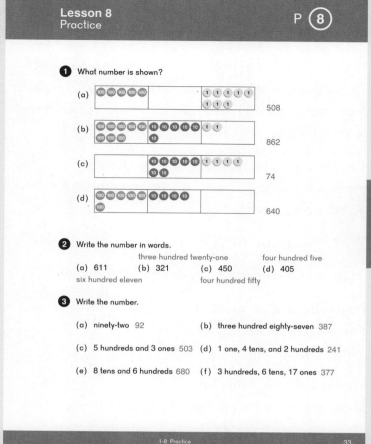

Lesson 8 Practice — P 8

1 What number is shown?
- (a) 508
- (b) 862
- (c) 74
- (d) 640

2 Write the number in words.
- (a) 611 — six hundred eleven
- (b) 321 — three hundred twenty-one
- (c) 450 — four hundred fifty
- (d) 405 — four hundred five

3 Write the number.
- (a) ninety-two — 92
- (b) three hundred eighty-seven — 387
- (c) 5 hundreds and 3 ones — 503
- (d) 1 one, 4 tens, and 2 hundreds — 241
- (e) 8 tens and 6 hundreds — 680
- (f) 3 hundreds, 6 tens, 17 ones — 377

1-8 Practice

33

▲ Cribbage

Materials: Cribbage board and pegs

This is a great age to teach students the game of Cribbage. Directions can be found online, or parents and grandparents can come into the classroom and teach students how to play.

The emphasis on mentally finding number combinations to 15 and 31 provides practice worth 1,000 worksheets!

▲ What Number Am I?

Here are some riddles to challenge students:

I am a three-digit number.
I have 24 ones and 21 tens.
What number am I?

★ I am a three-digit number.
My tens digit is 2 more than my hundreds digit.
My ones digit is 5 less than my tens digit.
The sum of my digits is 11.
What number am I?

Have students create their own and share!

Brain Works

★ **1,000**

Make 1,000 in many different ways. Examples:
900 + 50 + 50
500 + 300 + 150 + 50

Exercise 8 · page 23 ▶

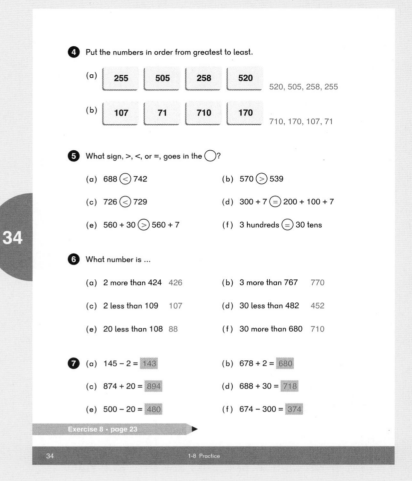

④ Put the numbers in order from greatest to least.

(a) | 255 | 505 | 258 | 520 | 520, 505, 258, 255

(b) | 107 | 71 | 710 | 170 | 710, 170, 107, 71

⑤ What sign, >, <, or =, goes in the ◯?

(a) 688 ⊂<⊃ 742   (b) 570 ⊂>⊃ 539

(c) 726 ⊂<⊃ 729   (d) 300 + 7 ⊂=⊃ 200 + 100 + 7

(e) 560 + 30 ⊂>⊃ 560 + 7   (f) 3 hundreds ⊂=⊃ 30 tens

⑥ What number is ...

(a) 2 more than 424   426   (b) 3 more than 767   770

(c) 2 less than 109   107   (d) 30 less than 482   452

(e) 20 less than 108   88   (f) 30 more than 680   710

⑦ (a) 145 − 2 = 143   (b) 678 + 2 = 680

(c) 874 + 20 = 894   (d) 688 + 30 = 718

(e) 500 − 20 = 480   (f) 674 − 300 = 374

Exercise 8 · page 23 ▶

34   1-8 Practice

34

Chapter 1 Numbers to 1,000

Exercise 1

Basics

❶ Fill in the missing numbers.

(a)

8 tens and [6] ones make 86.

[6] more than 80 is 86.

80 + [6] = 86

(b)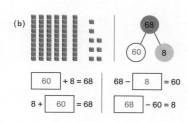

[60] + 8 = 68 | 68 − [8] = 60

8 + [60] = 68 | 68 − [60] = 8

(c) 90 and 7 make [97].

[90] more than 7 is 97.

[7] less than 97 is 90.

1-1 Tens and Ones   1

Practice

❷ (a) 40 + 9 = [49]   (b) 3 + 80 = [83]

60 + [2] = 62   [20] + 8 = 28

74 − 4 = [70]   51 − 50 = [1]

[95] − 5 = 90   [88] − 80 = 8

❸ Write the numbers.

(a) 4 tens and 8 ones   (b) seventy-two

[48]   [72]

(c) twenty-eight   (d) 7 ones and 3 tens

[28]   [37]

(e) four more than ninety   (f) three less than thirty-three

[94]   [30]

Challenge

❹ (a) 71 is [6] tens and 11 ones.

(b) 59 is 4 tens and [19] ones.

(c) 82 is 5 tens and [32] ones.

(d) 49 is 1 ten and [39] ones.

2   1-1 Tens and Ones

Exercise 2

Basics

❶ (a) Count on by ones and write the numbers between 35 and 43.

35	36	37	38	39	40	41	42	**43**

(b) Count back by ones and write the numbers between 83 and 75.

83	82	81	80	79	78	77	76	**75**

(c) Count on by tens and write the numbers between 5 and 85.

5	15	25	35	45	55	65	75	**85**

(d) Count back by tens and write the numbers between 99 and 19.

99	89	79	69	59	49	39	29	**19**

❷ (a) 2 more than 68 is [70].   2 less than 68 is [66].

20 more than 68 is [88].   20 less than 68 is [48].

(b) 3 more than 52 is [55].   3 less than 52 is [49].

30 more than 52 is [82].   30 less than 52 is [22].

1-2 Counting by Tens or Ones   3

Practice

❸ Complete the number patterns.

(a) | **73** | 75 | **77** | **79** | 81 | **83** |
|---|---|---|---|---|---|

(b) | **43** | 53 | **63** | **73** | 83 | **93** |
|---|---|---|---|---|---|

(c) | **73** | **76** | 79 | 82 | **85** | 88 |
|---|---|---|---|---|---|

(d) | 89 | **79** | 69 | 59 | **49** | **39** |
|---|---|---|---|---|---|

❹ (a) 82 + 2 = [84]   (b) 82 − 2 = [80]

(c) 58 + 3 = [61]   (d) 58 − 3 = [55]

(e) 42 + 20 = [62]   (f) 67 − 30 = [37]

❺ (a) 45 + [3] = 48   (b) 87 − [30] = 57

(c) [74] + 20 = 94   (d) [50] − 1 = 49

(e) [2] + 39 = 41   (f) [24] − 20 = 4

Challenge

❻ Sofia is putting together a picture album.
She puts 26 pictures in the album on Monday.
Each day, she adds 3 more pictures to the album.
How many pictures will be in the album after she
adds pictures on Thursday?   35 pictures

26 + 3 = 29 (Tues)
29 + 3 = 32 (Wed)
32 + 3 = 35 (Thu)

4   1-2 Counting by Tens or Ones

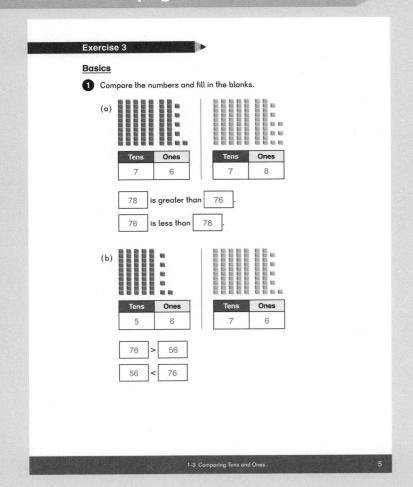

Exercise 3

Basics

1 Compare the numbers and fill in the blanks.

(a)

Tens	Ones
7	6

Tens	Ones
7	8

78 is greater than 76 .

76 is less than 78 .

(b)

Tens	Ones
5	6

Tens	Ones
7	6

76 > 56

56 < 76

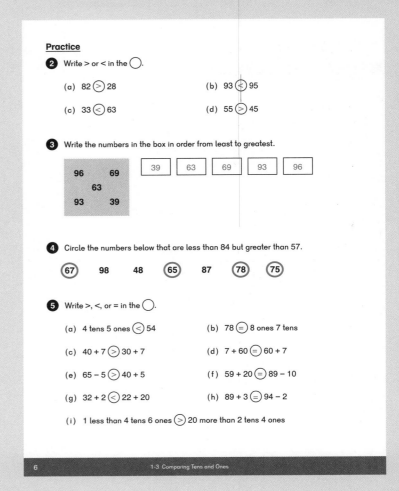

Practice

2 Write > or < in the ◯.

(a) 82 ⊚> 28   (b) 93 ⊚< 95

(c) 33 ⊚< 63   (d) 55 ⊚> 45

3 Write the numbers in the box in order from least to greatest.

96	69
	63
93	39

39 63 69 93 96

4 Circle the numbers below that are less than 84 but greater than 57.

(67) 98 48 (65) 87 (78) (75)

5 Write >, <, or = in the ◯.

(a) 4 tens 5 ones ⊚< 54   (b) 78 ⊚= 8 ones 7 tens

(c) 40 + 7 ⊚> 30 + 7   (d) 7 + 60 ⊚= 60 + 7

(e) 65 − 5 ⊚> 40 + 5   (f) 59 + 20 ⊚= 89 − 10

(g) 32 + 2 ⊚< 22 + 20   (h) 89 + 3 ⊚= 94 − 2

(i) 1 less than 4 tens 6 ones ⊚> 20 more than 2 tens 4 ones

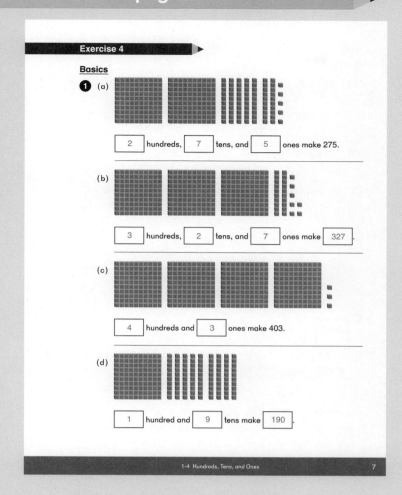

Exercise 4

Basics

1. (a) [2] hundreds, [7] tens, and [5] ones make 275.

(b) [3] hundreds, [2] tens, and [7] ones make [327].

(c) [4] hundreds and [3] ones make 403.

(d) [1] hundred and [9] tens make [190].

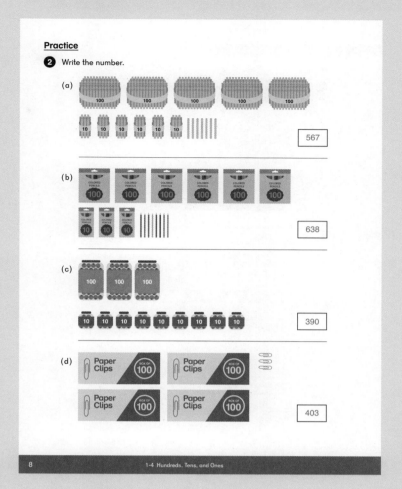

Practice

2. Write the number.

(a) [567]

(b) [638]

(c) [390]

(d) [403]

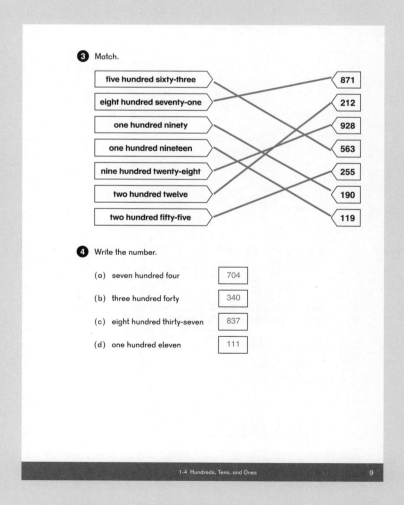

3. Match.

five hundred sixty-three	871
eight hundred seventy-one	212
one hundred ninety	928
one hundred nineteen	563
nine hundred twenty-eight	255
two hundred twelve	190
two hundred fifty-five	119

4. Write the number.

(a) seven hundred four — [704]

(b) three hundred forty — [340]

(c) eight hundred thirty-seven — [837]

(d) one hundred eleven — [111]

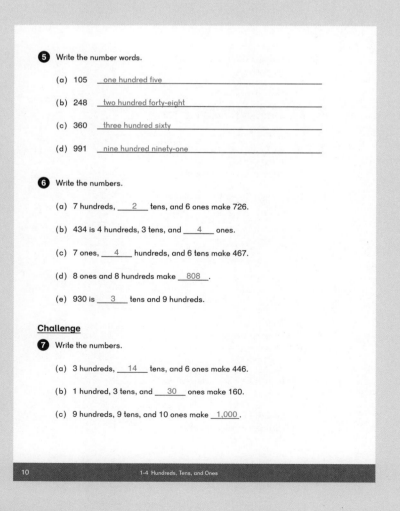

5. Write the number words.

(a) 105 — one hundred five

(b) 248 — two hundred forty-eight

(c) 360 — three hundred sixty

(d) 991 — nine hundred ninety-one

6. Write the numbers.

(a) 7 hundreds, __2__ tens, and 6 ones make 726.

(b) 434 is 4 hundreds, 3 tens, and __4__ ones.

(c) 7 ones, __4__ hundreds, and 6 tens make 467.

(d) 8 ones and 8 hundreds make __808__.

(e) 930 is __3__ tens and 9 hundreds.

Challenge

7. Write the numbers.

(a) 3 hundreds, __14__ tens, and 6 ones make 446.

(b) 1 hundred, 3 tens, and __30__ ones make 160.

(c) 9 hundreds, 9 tens, and 10 ones make __1,000__.

Exercise 5

Basics

1 Fill in the place-value chart with the number of hundreds, tens, and ones. Then write the number shown.

(a)
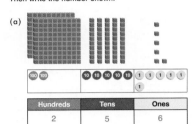

Hundreds	Tens	Ones
2	5	6

200 + 50 + 6 = 256

(b)

Hundreds	Tens	Ones
3	0	7

300 + 7 = 307

(c)

Hundreds	Tens	Ones
6	3	0

600 + 30 = 630

Practice

2 Write the number.

(a)

595

(b)
274

(c)
908

3 Color to show the number.   Students may color other discs to show the number.

(a) 109

(b) 780

(c) 400 + 40 + 5

4 Complete the chart.
Write the number.

7 + 80 + 300

Hundreds	Tens	Ones
3	8	7

387

The digit 3 is in the ___hundreds___ place.

5 Complete the chart.
Write the numbers.

4 hundreds, 8 ones

Hundreds	Tens	Ones
4	0	8

408

The digit __8__ is in the ones place.

The digit __0__ is in the tens place.

The digit __4__ is in the hundreds place.

6 (a) 700 + 60 + 5 = 765   (b) 400 + 6 = 406

(c) 200 + 20 + 2 = 222   (d) 800 + 70 = 870

(e) 30 + 6 + 500 = 536   (f) 60 + 800 = 860

(g) 10 + 3 + 900 = 913   (h) 7 + 700 = 707

7 Each number is a 3-digit number.
Write the number.

(a) The digit 4 is in the ones place.
    The digit in the hundreds place is 6.
    0 is in the tens place.   604

(b) The digit 7 is in the hundreds place.
    The digit 2 is in the ones place.
    The total of the digits is 13.   742

Challenge

8 (a) 10 ones = 1 ten   (b) 10 tens = 1 hundred

(c) 20 ones = 2 tens   (d) 30 tens = 3 hundreds

(e) 40 ones = 4 tens   (f) 60 tens = 6 hundreds

(g) 52 tens = 5 hundreds 2 tens

(h) 87 tens = 8 hundreds 7 tens

9 Color to show how many in all.   Students may color other discs to
Write the number.   show the number.

(a) 3 hundreds, 60 tens, 20 ones

920

(b) 8 hundreds, 2 tens, 63 ones
883

(c) 7 hundreds, 12 tens, 20 ones
840

(d) 1 hundred, 84 tens, 59 ones
999

Teacher's Guide 2A Chapter 1   © 2017 Singapore Math Inc.

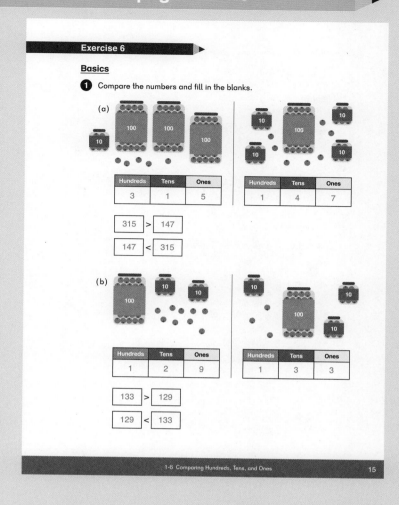

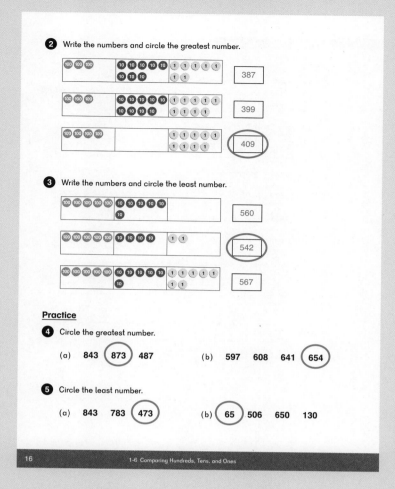

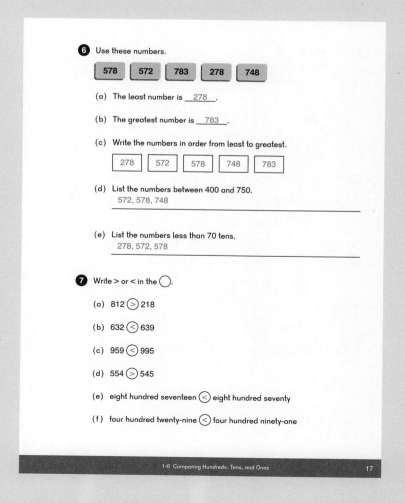

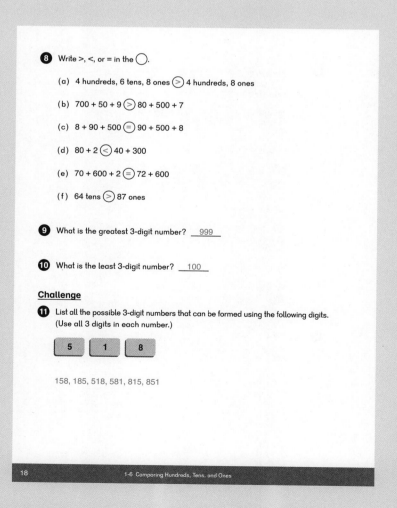

Exercise 7

Basics

1 (a) Count on by ones and write the numbers between 496 and 503.

496	497	498	499	500	501	502	**503**

(b) Count back by ones and write the numbers between 315 and 308.

315	314	313	312	311	310	309	**308**

(c) Count on by tens and write the numbers between 572 and 642.

572	582	592	602	612	622	632	**642**

(d) Count back by tens and write the numbers between 949 and 879.

949	939	929	919	909	899	889	**879**

(e) Count on by hundreds and write the numbers between 35 and 735.

35	135	235	335	435	535	635	**735**

(f) Count back by hundreds and write the numbers between 706 and 6.

706	606	506	406	306	206	106	**6**

1-7 Counting by Hundreds, Tens, or Ones   19

2 Color to show the number, then write the number.

 518

(a) 2 more than 518

 520

(b) 2 less than 518

 516

(c) 20 more than 518

 538

(d) 20 less than 518

 498

(e) 200 more than 518

 718

(f) 200 less than 518

718
318

20   1-7 Counting by Hundreds, Tens, or Ones

Practice

3 Color to show the number, then write the number.

(a) 200 more than 111

 311

(b) 3 less than 412

 409

(c) 10 more than 893

 903

(d) 20 less than 700

 680

4 (a) 824 + 2 = 826   (b) 320 − 10 = 310

(c) 518 + 10 = 528   (d) 367 − 300 = 67

(e) 420 + 100 = 520   (f) 67 − 30 = 37

(g) 20 + 890 = 910   (h) 200 − 1 = 199

1-7 Counting by Hundreds, Tens, or Ones   21

5 Complete the number patterns.

(a) | **234** | 334 | **434** | **534** | 634 | **734** |
(b) | **732** | **712** | 692 | 672 | **652** | **632** |
(c) | **230** | 233 | **236** | **239** | 242 | **245** |
(d) | 73 | **273** | **473** | 673 | 873 |

6 (a) 245 + 100 = 345   (b) 588 − 100 = 488

(c) 922 + 2 = 924   (d) 245 − 10 = 235

(e) 200 + 139 = 339   (f) 429 − 20 = 409

7 Write >, <, or = in the ◯.

(a) 310 + 20 < 210 + 200

(b) 480 − 1 < 480 + 20

(c) 630 + 100 = 930 − 200

(d) 298 + 3 > 308 − 20

22   1-7 Counting by Hundreds, Tens, or Ones

Exercise 8

Check

1 Write the number.

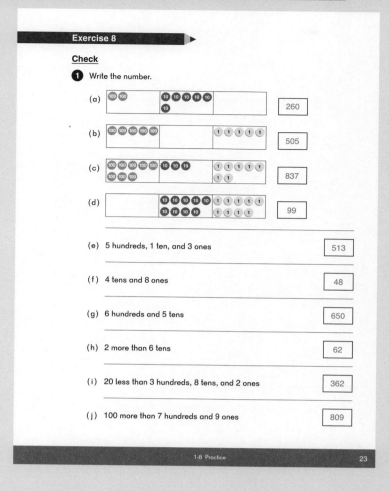

(a) 260

(b) 505

(c) 837

(d) 99

(e) 5 hundreds, 1 ten, and 3 ones — 513

(f) 4 tens and 8 ones — 48

(g) 6 hundreds and 5 tens — 650

(h) 2 more than 6 tens — 62

(i) 20 less than 3 hundreds, 8 tens, and 2 ones — 362

(j) 100 more than 7 hundreds and 9 ones — 809

2 Use the given digits to write the number.

(a) 2   8   6

The digit in the hundreds place is 2 less than the digit in the tens place.

682

(b) 6   9   1

The digit in the ones place is 3 more than the digit in the hundreds place.

619

3 (a) 80 + 9 = 89   (b) 66 − 6 = 60

(c) 500 + 63 = 563   (d) 7 + 200 = 207

(e) 899 + 2 = 901   (f) 680 − 10 = 670

(g) 508 + 100 = 608   (h) 1,000 − 300 = 700

4 Write >, <, or = in the ◯.

(a) 200 + 10 + 5 ⟨<⟩ 90 + 400 + 6

(b) 700 + 30 ⟨>⟩ 90 + 3

(c) 413 − 300 ⟨<⟩ 313 + 20

(d) 82 + 900 ⟨>⟩ 952 − 30

5 Circle the correct number.

(a) The number is greater than 300, less than 460, and has the digit 4 in the tens place.

244   412   **(348)**   364   590

(b) The number is greater than 6 hundreds and 9 tens, less than 80 + 700 + 2, and has the digit 0 in the ones place.

708   910   642   790   **(710)**

Challenge

6 Match the pattern to a number that will be in the pattern when it is continued on.

432, 434, 436, 438, ...   383

309, 409, 509, 609, ...   448

433, 423, 413, 403, ...   168

968, 768, 568, 368, ...   809

7 The numbers 1 to 1,000 are put in a number chart.
The first 6 rows are shown below.
Look for a pattern.

Column A	Column B	Column C	Column D	Column E
1	2	3	4	5
6	7	8	9	10
11	12	13	14	15
16	17	18	19	20
21	22	23	24	25
26	27	28	29	30

Write the rest of the numbers that will be in the row that starts with 861.

861	862	863	864	865

Write the letter of the column where the following numbers will be in the chart.

471   203   560   689   992

Column	A	C	E	D	B

Teacher's Guide 2A Chapter 1

Notes

Suggested number of class periods: 5–6

	Lesson	Page	Resources	Objectives
	Chapter Opener	p. 41	TB: p. 35	Practice adding and subtracting numbers to 20. Practice solving word problems.
1	Strategies for Addition	p. 43	TB: p. 37 WB: p. 27	Add with regrouping within 20 by making a 10.
2	Strategies for Subtraction	p. 45	TB: p. 39 WB: p. 29	Subtract with regrouping within 20 by decomposing the minuend or subtrahend.
3	Parts and Whole	p. 47	TB: p. 41 WB: p. 33	Understand the part-whole meanings of addition and subtraction. Solve problems involving a missing part or whole.
4	Comparison	p. 50	TB: p. 45 WB: p. 37	Solve addition and subtraction problems involving comparison.
5	Practice	p. 53	TB: p. 49 WB: p. 41	Practice adding and subtracting numbers to 20. Practice solving word problems.
	Workbook Solutions	p. 55		

In **Dimensions Math® 1A**, students learned to:

- Add and subtract within 20
- Solve basic word problems

In this chapter, students review their knowledge of basic addition and subtraction strategies. Numbers are kept within 20 to ensure students have mastery of numbers to 20 before **Chapter 3**: **Addition and Subtraction — Part 2**, where the vertical algorithms with regrouping are introduced.

As the first two lessons in this chapter are mainly review of strategies taught in 1A, students may progress quickly through them.

Bar Models

Lessons 3 and 4 introduce bar models. At this stage, students should be working with linking cubes and paper strips to solve the problems. They should not be expected to draw the bar models yet.

Because the problems involve numbers within 20, many students will solve the problems quickly and may only focus on the computation. Remind them that the problems are easy because they are learning a new strategy to solve problems and show their thinking. Challenge students who have mastered numbers to 20 by substituting greater numbers in the problems.

Models are included as a representational tool to help students understand the quantities in a word problem and their relationships, as well as understand what operations and processes to use to solve a problem.

Although students are not required to draw the models, teachers should focus students' attention on the parts of the model, the quantities, and their relationships to the words in the problem. In grade 3, students will be required to draw models and they will begin to see the model's usefulness gradually as they solve harder problems.

Bar models will be used for word problems throughout the **Dimensions Math®** series. There are two main types of models students will work with at this level: Part-Whole Models and Comparison Models.

Part-Whole Models

This type of model extends the understanding of part-whole relationships from a number bond to a more adaptable representation.

There are 10 students at the park. 7 students are boys. How many students are girls?

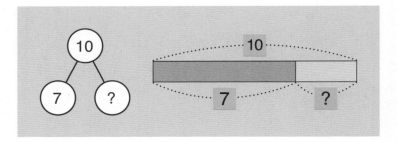

$10 - 7 = 3$
There are 3 girls at the park.

Students can easily see from the model, as they could from the number bond, that since they are given the whole and one part, they subtract to find the other part.

Comparison Models

This type of model allows students to compare quantities. They are particularly useful as number bonds do not portray comparison quite as well.

(a) How many more boys than girls are at the park?
(b) How many children are at the park altogether?

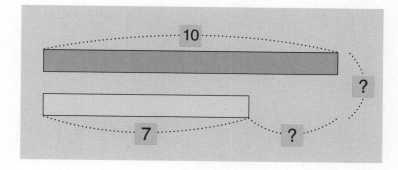

(a) 10 − 7 = 3
There are 3 more boys than girls at the park.
(b) 10 + 7 = 17
There are 17 children at the park altogether.

There are 10 boys and 7 girls at the park.
From this model, students can easily see that they need to subtract to find the answer to (a). Although (b) is a part-whole type of problem, the same model can be used to indicate this. If this representation confuses students, the two parts can be represented side-by-side instead, as is done on page 46 of the textbook.

Another advantage to bar models is that the bars, which represent numbers, can be moved around to show different relationships more clearly. This will become more important in later levels as the word problems become more complex.

For word problems, students should write an equation and an answer sentence. It is better to write the answer sentence after solving the problem. Writing the answer in a complete sentence will encourage students to re-read the problem and determine whether they have answered the correct question in the problem. This will also become more important in multi-step word problems where students often solve the first step and believe they are done with the problem.

Materials

- Linking cubes
- Game board from the classroom
- Art paper
- Playing cards
- Index cards or construction paper
- Markers or crayons
- Paper strips in 2 colors
- Play money

Blackline Masters

- Addition and Subtraction to 10 Fact Cards
- Alligator Cards
- Number Cards

Storybooks

- *The Mission of Addition* by Brian P. Cleary
- *Mission: Addition* by Loreen Leedy
- *12 Ways to Get to 11* by Eve Merriam
- *What's New at the Zoo?* by Suzanne Slade

Activities

Games and activities included in this chapter are designed to provide practice and extensions of adding and subtracting within 20. They can be used after students complete the **Do** questions, or anytime review and practice are needed. Students should know these facts from memory by the end of the chapter.

Objectives

- Practice adding and subtracting numbers to 20.
- Practice solving word problems.

This **Chapter Opener** is designed as a review of basic addition and subtraction facts. Students should be assessed on their mastery of facts and should make flash cards for any facts they still need to practice.

By the end of the chapter, students should know all facts to 20 from memory.

Students who do not know the facts from memory should continue to practice them until they are mastered. As such, multiple activities are included in the lessons for ongoing practice.

Students should notice a pattern with the addition fact cards: If either addend increases, the answer increases by the same amount.

Patterns students should notice with the subtraction fact cards include:

- If the first number increases, the answer increases by the same amount.
- If the second number increases, the answer decreases.

Ask students, "Why is this so?" and, "Why is it different from addition?"

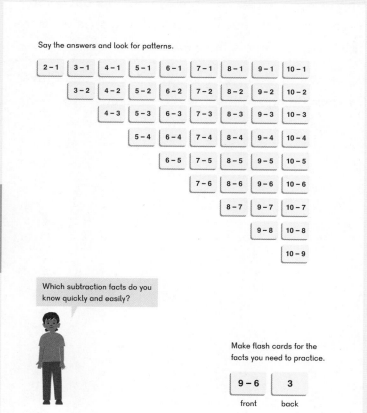

Activities

▲ Takeover!

Materials: A game board from the classroom, Addition and Subtraction to 10 Fact Cards (BLM)

Almost any board game can be "taken over" for some subtraction practice. Many board games require the use of dice to determine how many squares to move ahead. In **Takeover!**, players change the rules for moving ahead. Instead of dice, use a deck of subtraction flash cards.

On each turn, the player draws a flash card and figures out the answer. For example, if a player draws 7 − 2, he moves forward 5 spaces.

★ Extend by having students create their own boards.

▲ Rock, Paper, Scissors, Math!

Players bounce fists in their hand while saying, "Rock, Paper, Scissors, Math."

On the word "math," each player shoots out some fingers on one hand. The student who says the sum of the fingers first is the winner.

For example, if Player 1 shows 7 fingers and Player 2 shows 6 fingers, the first player to say, "13," wins.

▲ Alligator! Alligator! Alligator!

Materials: Alligator Cards (BLM), Addition and Subtraction to 10 Fact Cards (BLM)

Spread some of the cards facedown in front of the group of students. Include at least 3 Alligator Cards (BLM). Students take turns selecting a card. The student reads the expression on the card and gives the answer.

When an alligator comes up, all the students shout, "Alligator! Alligator! Alligator!"

While the game seems goofy and may be loud, it keeps all students engaged because they do not want to miss the alligator card and the opportunity to shout, "Alligator! Alligator! Alligator!" when an alligator card is selected.

Collect the used cards as they are called.

Objective

- Add with regrouping within 20 by making a 10.

Lesson Materials

- 20 linking cubes, 10 each of 2 different colors per student

Think

Provide students with linking cubes and pose the **Think** problem. Have students share their equation and how they solved the problem.

Learn

Have students discuss the strategies that Dion and Sofia used. Ask students how many solved their problem like Dion, and how many like Sofia.

Both Dion and Sofia start by making 10, a strategy that students learned in **Dimensions Math® 1A** and used throughout **Dimensions Math® 1B**.

Dion decomposes, or splits, the 7 into 2 and 5. He adds the 8 and the 2 to get 10, then easily adds the remaining 5. 8 + 2 = 10, 10 + 5 = 15

Sofia splits the 8 into 5 and 3. 3 + 7 = 10, 10 + 5 = 15

Students should see that these are two ways to solve the same problem, and that they are not looking at two different problems.

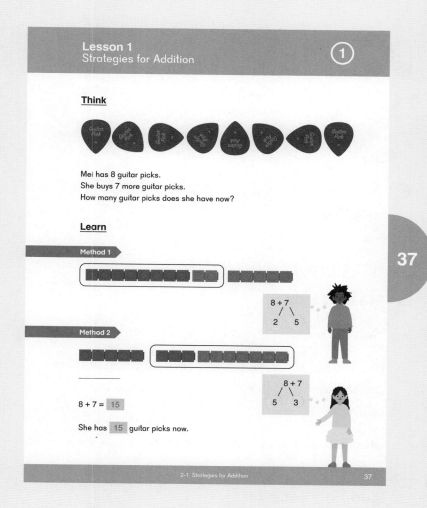

Do

1 Ask students why Emma split the second number in her problem while Mei split the first number. Students may note that the 9 only needs 1 more to make a 10. Some students may find that splitting 9 into 5 and 4 is just as efficient.

2 Have students discuss both ways of splitting the addends.

Activities

▲ Flash Cards

Materials: Index cards or construction paper

Provide students with index cards and have them create their own flash cards for future practice and games.

Students can also fold construction paper into 8 equal parts and cut out their own flash cards.

▲ Addition Face-off

Materials: Number Cards (BLM) 0 to 10, or regular playing cards

Play in groups of 2—4. If using a regular deck of cards, aces are one, and face cards are ten.

Deal out cards evenly to players. Players flip over 2 cards each and call out their sums. The student with the greatest sum wins and collects all the cards.

If there is a tie, repeat, turning over 2 more cards.

Exercise 1 • page 27

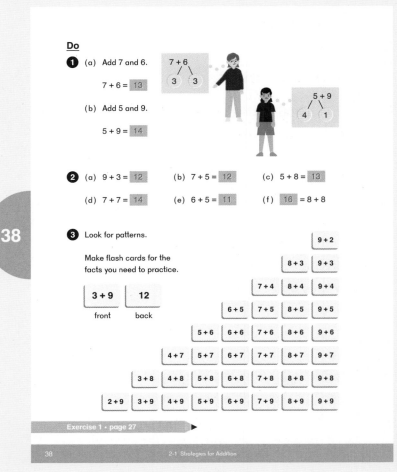

Lesson 2 Strategies for Subtraction

Objective

- Subtract with regrouping within 20 by decomposing the minuend or subtrahend.

Lesson Materials

- 20 linking cubes, 10 each of 2 different colors for each student

Think

Pose the **Think** problem. Students should represent Alex's card problem with linking cubes. Have them write the equation, find the answer, and then share how they solved the problem.

If needed, prompt students to recall methods they learned from a previous grade level. Ask, "Can we split 15 or split 7 to make it easy to subtract by using the facts within 10?"

Learn

Mei's strategy is to split the minuend (first number or whole) into 10 and 5 ones. She is subtracting 7 from 10. In grade 1, this is referred to as "subtracting from ten." $10 - 7 = 3$, $3 + 5 = 8$

Emma's strategy is to split the subtrahend (second number or part). First she subtracts 5 from 15 to get 10, then she subtracts a further 2 from 10 to get her answer. In grade 1, this is referred to as "double subtracting" or "subtract twice." $15 - 5 = 10$, $10 - 2 = 8$

Ask students if they solved their problem like Mei or like Emma. As with the addition example in the previous lesson, students should see that these are two different ways to solve the same problem.

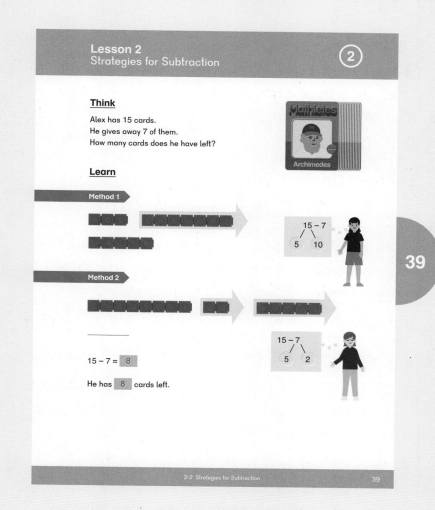

Do

2 Have students share and discuss how they split the numbers in the problems.

Activities

▲ Flash Cards

Materials: Index cards or construction paper

As in the previous lesson, provide students with index cards or construction paper, and have them create their own flash cards for future practice.

▲ Salute!

Materials: Deck of cards with face cards removed

Salute! is played with three students. The Caller shuffles, then deals out the deck to two players. The third player is the Caller.

When the Caller says, "Salute!" the other players place the top cards from their piles on their foreheads. The two players can see each other's cards, but not their own.

The Caller tells the players the sum of the two numbers on their cards. (Think of the three players as a number bond with one of the addends missing.)

The player who says the number on his card first is the winner.

The sum is 13.

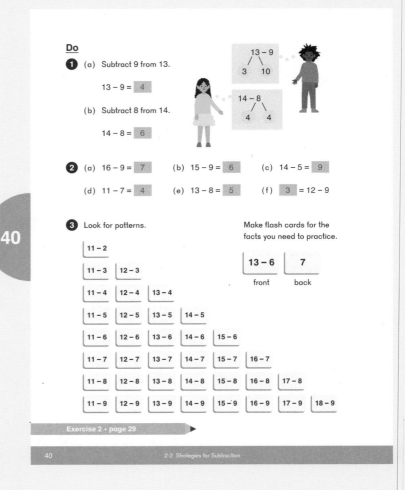

▲ Alligator! Alligator! Alligator!

Materials: Alligator Cards (BLM), Addition and Subtraction to 20 Fact Cards (BLM)

Play the game from the **Chapter Opener** on page 42 of this Teacher's Guide, but with fact cards to 20.

Exercise 2 • page 29

Lesson 3 Parts and Whole

Objectives

- Understand the part-whole meanings of addition and subtraction.
- Solve problems involving a missing part or whole.

Lesson Materials

- 20 linking cubes, 10 each of 2 different colors for each student
- Paper strips, 2 colors of the same length

Think

Read the two **Think** stories aloud and have students compare and discuss the two different stories. Ask students how the two stories are the same or how they are different.

Have students draw the number bonds for the two problems, then write the equations and find the answers.

Have students discuss why they added or subtracted.

Learn

In this and the next lesson, bar models are introduced as a way to help students visualize the problem situation. Provide students with linking cubes and have them model each story. Students are not required to draw the models.

Alex thinks that if he knows two parts in a problem, he can add them together to find the whole, just like in a number bond.

Sofia's number bond shows that if she knows a whole and one part, then she can find the missing part. She will use subtraction to find the missing part.

Tell students these are called **part-whole** bar models. If we know the two parts, we can find the whole. If we know a whole and a part, we can find the missing part.

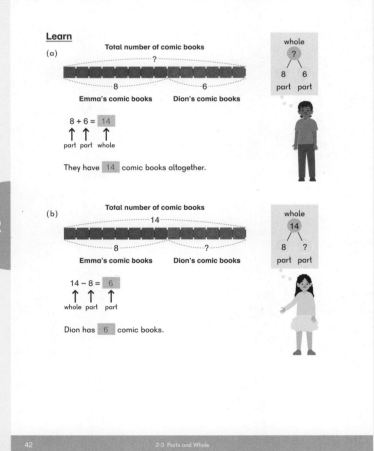

Do

1 Students can find four facts from the bar model just as they have from the number bond.

2 Ask students:

- What is similar and what is different about the model in this problem compared to the cubes in the previous problems?
- What does the question mark mean?

In this problem, the bar model diagram changes from individual cubes that can be counted to a length that is somewhat proportional. Students can continue to use linking cubes as needed.

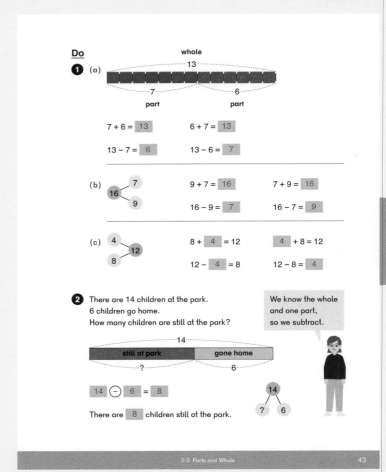

3 Provide students with paper strips. Have them represent the parts of the bar model with the paper strips on a whiteboard, where they can write the numbers above and below.

Students should begin with two strips the same length.

Ask:

• Which strip will represent the children?
• Which will represent the adults?
• Are there the same number of children and adults?
• Should our strips be the same length?

Have students fold the strip that represents the adults to be a bit shorter than the strip that represents the children.

Have them put the strips together, as shown in the textbook, to find the whole.

4 Paper strips may still be used to find the difference, however, students do not have experience estimating to find the missing part. Have them start with strips of equal length and fold them after completing the problem.

Have students keep their strips for future lessons.

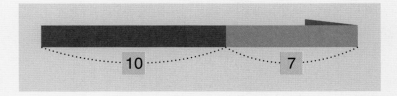

Exercise 3 • page 33

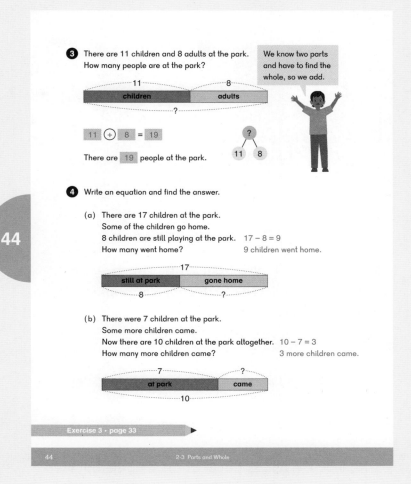

3 There are 11 children and 8 adults at the park. How many people are at the park?

We know two parts and have to find the whole, so we add.

11 + 8 = 19

There are 19 people at the park.

4 Write an equation and find the answer.

(a) There are 17 children at the park.
    Some of the children go home.
    8 children are still playing at the park. 17 – 8 = 9
    How many went home?   9 children went home.

(b) There were 7 children at the park.
    Some more children came.
    Now there are 10 children at the park altogether. 10 – 7 = 3
    How many more children came?   3 more children came.

Exercise 3 • page 33

44   2-3 Parts and Whole

Lesson 4 Comparison

Objective

- Solve addition and subtraction problems involving comparison.

Lesson Materials

- 20 linking cubes, 10 each of 2 different colors for each student
- Paper strips, 2 colors of the same length

Think

Pose the two problems from **Think**. Have students represent the quantities with cubes or paper strips to solve the problem.

Ask:

- How can we use what we learned in the last lesson?
- Can we use our cubes or paper strips to model this problem?
- How do the cubes/paper strips help determine what equation to write?

Have students share their strategies.

Learn

Have students compare and discuss the two different stories and equations:

- How are the linking cubes lined up in this example?
- Why do you think they are lined up that way?
- Where are the "more" ants represented on the cubes?
- Where are "all of the insects represented" on the cubes?
- What do the 2 question marks represent?
- Why is one of them written on the right side?
- How can we find how many more ants there are than crickets?
- How can we find how many insects there are altogether?

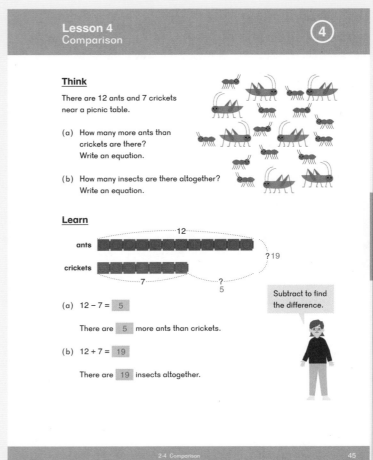

Emma tells us that to find the difference numerically, we can subtract.

Tell students these are called **comparison** bar models. We can use them to tell the difference between two or more quantities.

Do

1 Students can continue to use linking cubes or paper strips.

If students use paper strips, they should begin with two strips the same length.

Ask:

* Which strip will represent the ducks?
* Which will represent the swans?

1 (b) Have students stack the strips, one above the other, as shown in the textbook. Explain that the problem is comparing the number of ducks to the number of swans. When comparing, we want to know which quantity is more and which is less.

While there are four different equations that can be found from this model, they are not a fact family. Have students come up with a story for each equation. For example:

* 11 − 5 = ? There are 11 chairs and 5 tables. How many more chairs? How many fewer tables?
* 11 − 6 = ? There are 11 chairs. There are 6 more chairs than tables (or 6 fewer tables than chairs). How many tables?
* 5 + 6 = ? There are 5 tables. There are 6 more chairs than tables (or 6 fewer tables than chairs). How many chairs?
* 11 + 5 = ? There are 11 chairs and 5 tables. How many pieces of furniture are there in all?

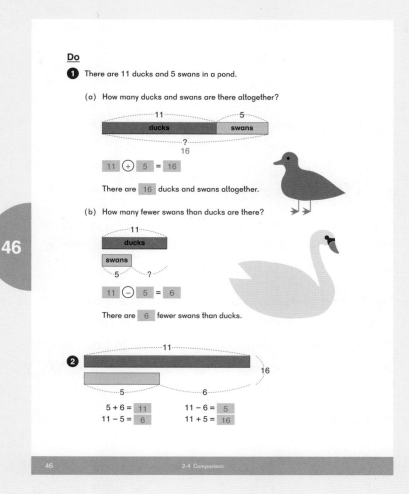

46

3 — 8 Ask students questions about the problems:

- What do we need to find?
- Are we comparing two quantities?
- Do we know the whole or total?
- Do we know one part?
- Where is the more/fewer in the model?
- How do we know whether to subtract or add?
- In **5**, which question mark represents the total?

Students should keep their paper strips as a tool for use in the future, or keep extra paper strips with other math tools in the classroom.

By the end of this lesson, many students may not need the paper strips or linking cubes to represent the problems. Students who are interested and prefer to draw the bar models should be allowed to do so.

Activity

▲ Model Posters

Materials: Paper strips (2 colors of the same length), markers or crayons, art paper, glue

Provide students with paper strips and an equation such as 9 + 3 = ☐ or 14 − ☐ = 7. Have them:

- Create a word problem.
- Illustrate the problem.
- Glue down the strips for the bar model for the problem.
- Write the equation and solution.

Have students discuss and share their work.

★ For a greater challenge, have students write their own equations.

Exercise 4 · page 37

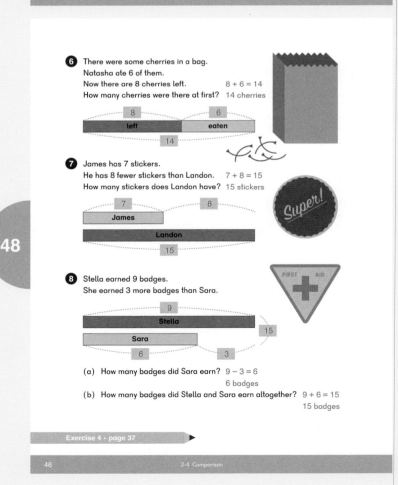

Write an equation and find the answer.

3 Kaylee is 7 years old.
Her sister Madison is 4 years older than Kaylee.
How old is Madison?

7 + 4 = 11
11 years old

4 Hudson has 18 stickers.
He has 5 more stickers than Darryl.
How many stickers does Darryl have?

18 − 5 = 13
13 stickers.

5 Jordon has 10 sports cards.
He has 4 fewer sports cards than Mila.

(a) How many cards does Mila have?
10 + 4 = 14
14 cards.

(b) How many cards do they have altogether?
10 + 14 = 24
24 cards.

2-4 Comparison   47

6 There were some cherries in a bag.
Natasha ate 6 of them.
Now there are 8 cherries left.
How many cherries were there at first?
8 + 6 = 14   14 cherries

7 James has 7 stickers.
He has 8 fewer stickers than Landon.
How many stickers does Landon have?
7 + 8 = 15   15 stickers

8 Stella earned 9 badges.
She earned 3 more badges than Sara.

(a) How many badges did Sara earn? 9 − 3 = 6
6 badges

(b) How many badges did Stella and Sara earn altogether? 9 + 6 = 15
15 badges

Exercise 4 · page 37

48   2-4 Comparison

Objectives

- Practice adding and subtracting numbers to 20.
- Practice solving word problems.

Practice

After students complete the **Practice** in the textbook, have them continue adding numbers to 20 by playing games from this chapter.

Students should be fluent with these facts prior to **Chapter 3: Addition and Subtraction — Part 2**.

2 – 8 Allow students to use their strips or linking cubes as needed.

Activity

★ **KenKen**

Easy **KenKen** activities are a great way to practice both addition and subtraction as well as logical thinking.

Some websites allow activities to be customized before printing to use addition only, or addition and subtraction, for younger students. Search the web for:

- kenkenpuzzle.com
- mathdoku.com
- calcudoku.org

An example **KenKen** board is shown to the right.

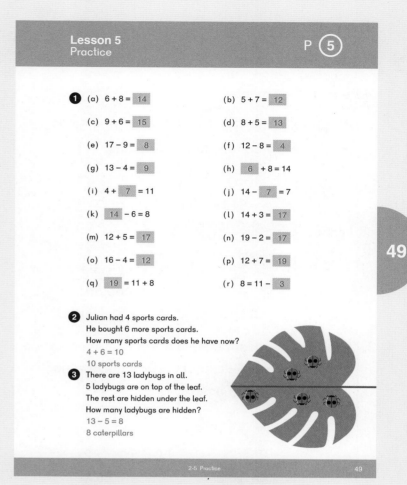

Lesson 5 Practice   P 5

1
(a) 6 + 8 = 14   (b) 5 + 7 = 12
(c) 9 + 6 = 15   (d) 8 + 5 = 13
(e) 17 – 9 = 8   (f) 12 – 8 = 4
(g) 13 – 4 = 9   (h) 6 + 8 = 14
(i) 4 + 7 = 11   (j) 14 – 7 = 7
(k) 14 – 6 = 8   (l) 14 + 3 = 17
(m) 12 + 5 = 17   (n) 19 – 2 = 17
(o) 16 – 4 = 12   (p) 12 + 7 = 19
(q) 19 = 11 + 8   (r) 8 = 11 – 3

2 Julian had 4 sports cards.
He bought 6 more sports cards.
How many sports cards does he have now?
4 + 6 = 10
10 sports cards

3 There are 13 ladybugs in all.
5 ladybugs are on top of the leaf.
The rest are hidden under the leaf.
How many ladybugs are hidden?
13 – 5 = 8
8 caterpillars

2-5 Practice   49

49

Brain Works

★ Money Problems

Materials: Play money

1. Grace has $13. She has $7 more than Eli. How much money do they have altogether?

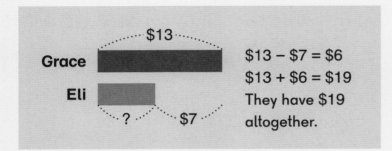

Grace
Eli

$13 − $7 = $6
$13 + $6 = $19
They have $19 altogether.

2. Grace, Eli and Isaac have $18 altogether. Grace has $9. Eli has $2 less than Grace. How much money does Isaac have?

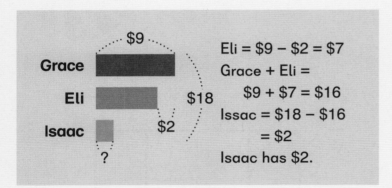

Grace
Eli
Isaac

Eli = $9 − $2 = $7
Grace + Eli =
$9 + $7 = $16
Issac = $18 − $16
= $2
Isaac has $2.

3. Grace has $6 more than Eli. After she gives some money to Eli they have the same amount. How much did money did she give to Eli?

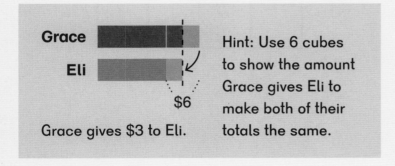

Grace
Eli

Hint: Use 6 cubes to show the amount Grace gives Eli to make both of their totals the same.

Grace gives $3 to Eli.

Students could use money and bar models to work through the problems.

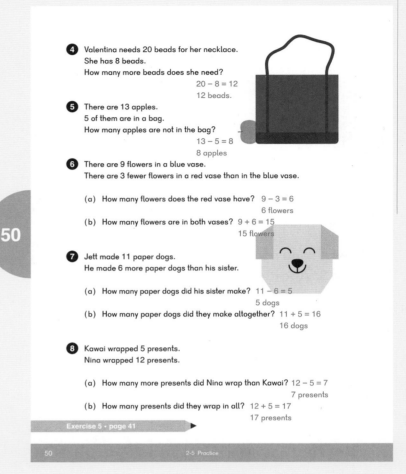

4   Valentina needs 20 beads for her necklace.
She has 8 beads.
How many more beads does she need?
20 − 8 = 12
12 beads.

5   There are 13 apples.
5 of them are in a bag.
How many apples are not in the bag?
13 − 5 = 8
8 apples

6   There are 9 flowers in a blue vase.
There are 3 fewer flowers in a red vase than in the blue vase.

(a) How many flowers does the red vase have?   9 − 3 = 6
6 flowers

(b) How many flowers are in both vases?   9 + 6 = 15
15 flowers

7   Jett made 11 paper dogs.
He made 6 more paper dogs than his sister.

(a) How many paper dogs did his sister make?   11 − 6 = 5
5 dogs

(b) How many paper dogs did they make altogether?   11 + 5 = 16
16 dogs

8   Kawai wrapped 5 presents.
Nina wrapped 12 presents.

(a) How many more presents did Nina wrap than Kawai?   12 − 5 = 7
7 presents

(b) How many presents did they wrap in all?   12 + 5 = 17
17 presents

Exercise 5 · page 41

50   2-5 Practice

Exercise 5 • page 41

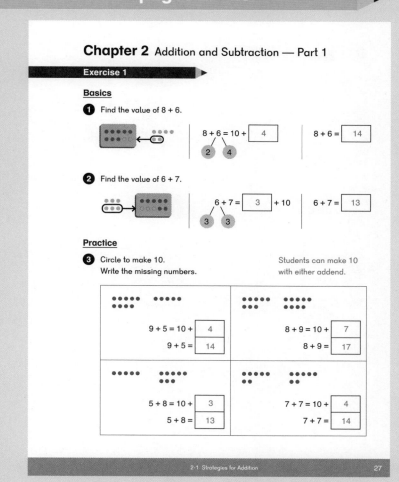

Chapter 2 Addition and Subtraction — Part 1

Exercise 1

Basics

1 Find the value of 8 + 6.

8 + 6 = 10 + [4]

2   4

8 + 6 = [14]

2 Find the value of 6 + 7.

6 + 7 = [3] + 10

3   3

6 + 7 = [13]

Practice

3 Circle to make 10.
Write the missing numbers.

Students can make 10 with either addend.

9 + 5 = 10 + [4]
9 + 5 = [14]

8 + 9 = 10 + [7]
8 + 9 = [17]

5 + 8 = 10 + [3]
5 + 8 = [13]

7 + 7 = 10 + [4]
7 + 7 = [14]

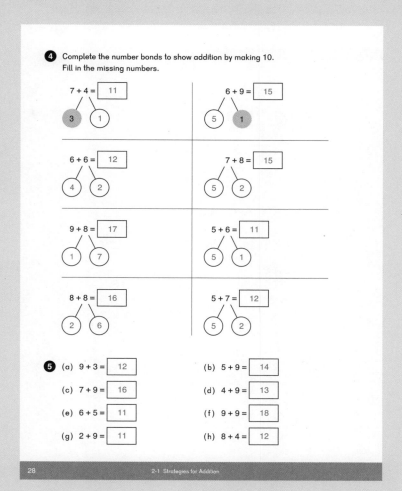

4 Complete the number bonds to show addition by making 10.
Fill in the missing numbers.

7 + 4 = [11]
3   1

6 + 9 = [15]
5   1

6 + 6 = [12]
4   2

7 + 8 = [15]
5   2

9 + 8 = [17]
1   7

5 + 6 = [11]
5   1

8 + 8 = [16]
2   6

5 + 7 = [12]
5   2

5 (a) 9 + 3 = [12]   (b) 5 + 9 = [14]

(c) 7 + 9 = [16]   (d) 4 + 9 = [13]

(e) 6 + 5 = [11]   (f) 9 + 9 = [18]

(g) 2 + 9 = [11]   (h) 8 + 4 = [12]

Exercise 2

Basics

1 Find the value of 14 − 8 by subtracting from the ten.

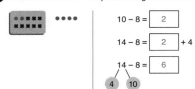

10 − 8 = 2

14 − 8 = 2 + 4

14 − 8 = 6

4   10

2 Cross off on the ten-frame card.

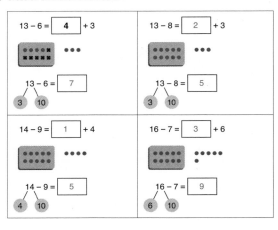

13 − 6 = **4** + 3

13 − 6 = 7

3   10

13 − 8 = 2 + 3

13 − 8 = 5

3   10

14 − 9 = 1 + 4

14 − 9 = 5

4   10

16 − 7 = 3 + 6

16 − 7 = 9

6   10

3 Find the value of 12 − 8 by subtracting the ones, then more ones.

12 − 2 = 10

10 − 6 = 4

12 − 8 = 4

8
2   6

4 Cross off the ones first.

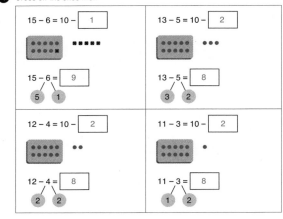

15 − 6 = 10 − 1

15 − 6 = 9

5   1

13 − 5 = 10 − 2

13 − 5 = 8

3   2

12 − 4 = 10 − 2

12 − 4 = 8

2   2

11 − 3 = 10 − 2

11 − 3 = 8

1   2

Practice

5 Complete the number bonds.
Fill in the missing numbers.

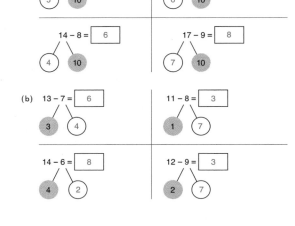

(a)   15 − 9 = 6

5   10

16 − 7 = 9

6   10

14 − 8 = 6

4   10

17 − 9 = 8

7   10

(b)   13 − 7 = 6

3   4

11 − 8 = 3

1   7

14 − 6 = 8

4   2

12 − 9 = 3

2   7

6 (a) 11 − 9 = 2

(b) 13 − 7 = 6

(c) 16 − 8 = 8

(d) 12 − 5 = 7

(e) 13 − 4 = 9

(f) 17 − 8 = 9

(g) 11 − 6 = 5

(h) 13 − 9 = 4

7 Add or subtract the ones.

(a) 5 + 3 = 8

15 + 3 = 18

10   5

(b) 9 − 3 = 6

19 − 3 = 16

10   9

8 (a) 14 + 2 = 16

(b) 16 − 4 = 12

(c) 6 + 8 = 14

(d) 18 − 9 = 9

(e) 9 + 6 = 15

(f) 12 − 7 = 5

(g) 12 + 5 = 17

(h) 17 − 4 = 13

(i) 4 + 8 = 12

(j) 14 − 5 = 9

Challenge

9 The numbers at the corners of the triangles
add up to the number between them.
Find the missing numbers.

(a)
4
11   10
7   13   6

(b)
9
15   16
6   13   7

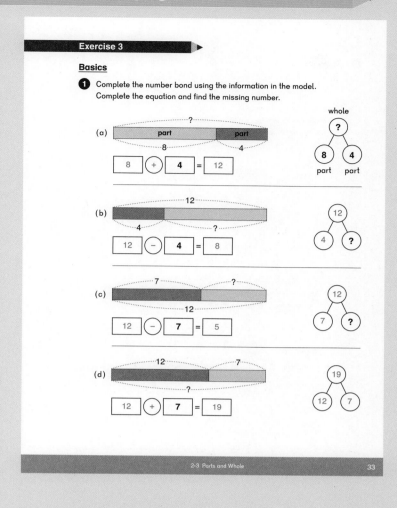

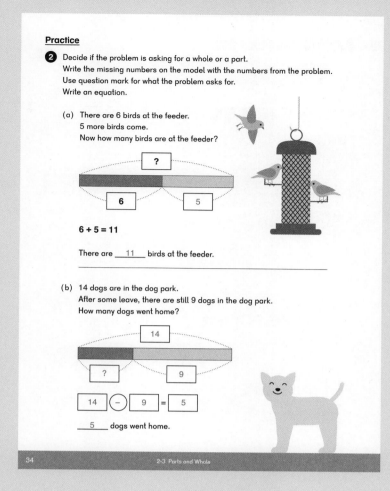

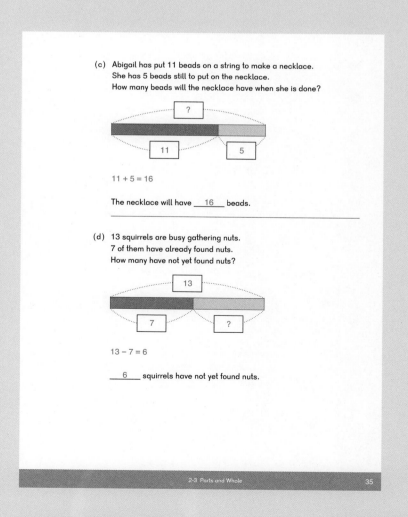

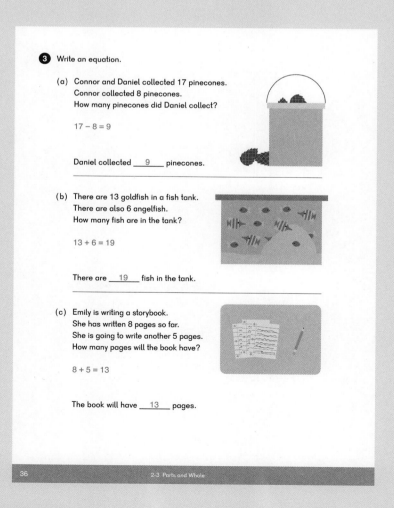

© 2017 Singapore Math Inc.   Teacher's Guide 2A Chapter 2   57

Exercise 4

Basics

1 Use the information in the model and complete the equation.

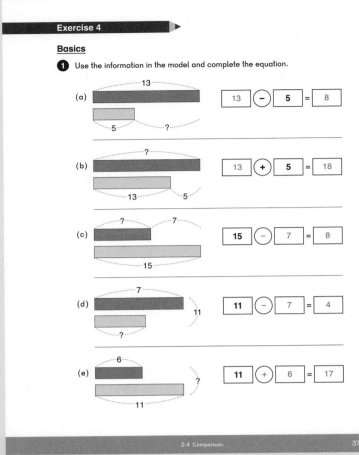

(a) $13 - 5 = 8$

(b) $13 + 5 = 18$

(c) $15 - 7 = 8$

(d) $11 - 7 = 4$

(e) $11 + 6 = 17$

Practice

2 Complete the models with the information in the problem.
Use a question mark for what the problem asks for.
Write an equation.

(a) There are 7 bulldozers at a construction site.
There are 4 more loaders than bulldozers at the site.
How many loaders are there?

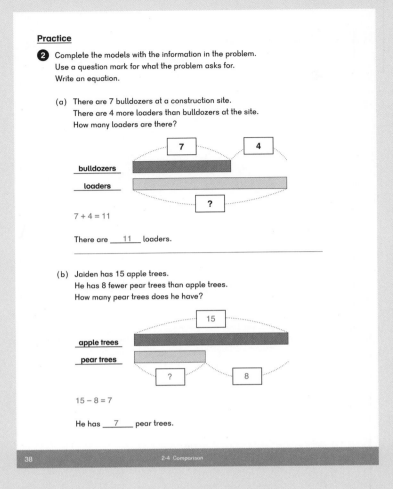

$7 + 4 = 11$

There are ___11___ loaders.

(b) Jaiden has 15 apple trees.
He has 8 fewer pear trees than apple trees.
How many pear trees does he have?

$15 - 8 = 7$

He has ___7___ pear trees.

(c) There are 9 swans at the pond.
There are 4 fewer swans than ducks.
How many ducks are there?

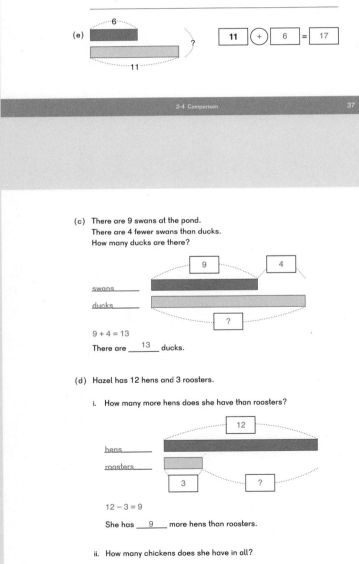

$9 + 4 = 13$

There are ___13___ ducks.

(d) Hazel has 12 hens and 3 roosters.

i. How many more hens does she have than roosters?

$12 - 3 = 9$

She has ___9___ more hens than roosters.

ii. How many chickens does she have in all?

$12 + 3 = 15$

She has ___15___ chickens in all.

3 Write an equation.

(a) Mr. Jackson planted 11 rosebushes.
He planted 4 more rosebushes than holly bushes.
How many holly bushes did he plant?

$11 - 4 = 7$

He planted ___7___ holly bushes.

(b) Hazel's hens laid 12 white eggs and 9 brown eggs.
How many fewer brown eggs did the hens lay than white eggs?

$12 - 9 = 3$

The hens laid ___3___ fewer brown than white eggs.

Challenge

4 Debra made some apple pies and cherry pies.
She made 10 apple pies.
She made 3 fewer cherry pies than apple pies.
How many pies did she make in all?

$10 - 3 = 7$
$10 + 7 = 17$

She made ___17___ pies.

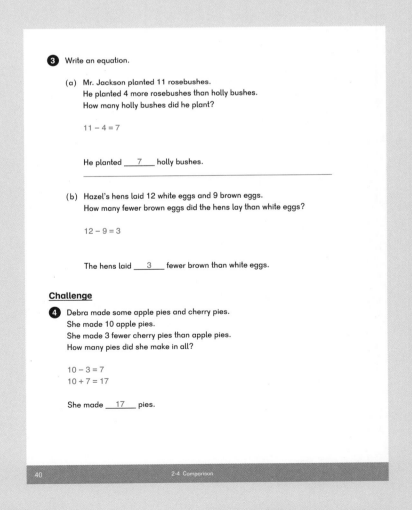

Exercise 5

<u>Check</u>

1 Complete the addition tables.

+	6	8	5	1	9	2	7	4	3
7	13	15	12	8	16	9	14	11	10
9	15	17	14	10	18	11	16	13	12
8	14	16	13	9	17	10	15	12	11
6	12	14	11	7	15	8	13	10	9

2 (a) $\boxed{6}$ + 8 = 14   (b) 12 − $\boxed{5}$ = 7

(c) $\boxed{15}$ − 9 = 6   (d) 4 = $\boxed{11}$ − 7

(e) 5 + $\boxed{9}$ = 14   (f) 15 = 8 + $\boxed{7}$

(g) $\boxed{4}$ + 11 = 15   (h) $\boxed{17}$ − 3 = 14

3 Write >, <, or = in the ◯.

(a) 15 − 7 ⟮ > ⟯ 11 − 9   (b) 8 + 3 ⟮ = ⟯ 5 + 6

(c) 12 − 9 ⟮ = ⟯ 14 − 11   (d) 9 + 9 ⟮ < ⟯ 21 − 2

(e) 17 − 8 ⟮ > ⟯ 87 − 80   (f) 215 − 200 ⟮ = ⟯ 7 + 8

4 Follow the arrows and fill in the missing numbers.
Write + or − in the ◯.

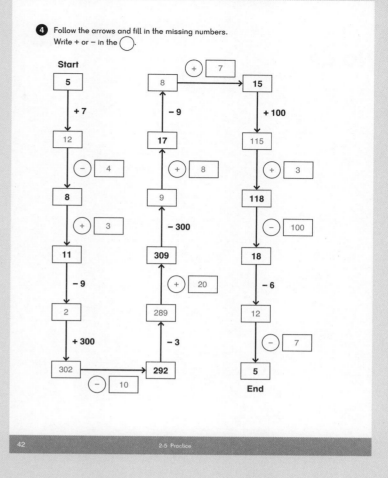

5 There are 17 cars in the parking lot.
After some cars drive away, 5 cars are still in the parking lot.
How many cars drove away?

17 − 5 = 12

_12_ cars drove away.

6 Kaylee folded 12 paper birds and 8 paper cats.
How many fewer paper cats did she fold than paper birds?

12 − 8 = 4

She folded _4_ fewer paper cats than paper birds.

7 There are 11 roses in a vase.
There are 5 more carnations than roses in the vase.
How many carnations are in the vase?

11 + 5 = 16

There are _16_ carnations.

8 There are 11 hamsters in a pet store.
There are 5 more hamsters than gerbils in the pet store.
How many gerbils are in the pet store?

11 − 5 = 6

There are _6_ gerbils.

<u>Challenge</u>

9 Jack has 3 more action figures than Jade.
They each get 6 more action figures.
How many more action figures does Jack have than Jade?

> If they each get the same amount, the difference does not change. Students can use counters if necessary.

Jack has _3_ more action figures than Jade.

10 Emiliano scored 15 points in a game.
Dexter scored 9 fewer points then Emiliano.
Cora scored 11 more points than Dexter.
How many more points did Cora score than Emiliano?

15 − 9 = 6 (Dexter's score)
6 + 11 = 17 (Cora's score)
17 − 15 = 2

> There are various ways to solve this. One method is the 3 equations shown. The other is to notice that 9 less and 11 more is 2 more.

Cora scored _2_ more points than Emiliano.

11 Complete the addition tables.

(a)
+	7	9	6
6	13	15	12
5	12	14	11
8	15	17	14

(b)
+	9	7	8
3	12	10	11
5	14	12	13
4	13	11	12

If students have trouble, suggest they simply start by filling in whatever can be filled in. That will allow other squares to be filled in.

Notes

Suggested number of class periods: 12–13

	Lesson	Page	Resources		Objectives
	Chapter Opener	p. 65	TB:	p. 51	Investigate addition of three-digit numbers.
1	Addition Without Regrouping	p. 66	TB: WB:	p. 52 p. 45	Add within 1,000 without regrouping by using an algorithm.
2	Subtraction Without Regrouping	p. 69	TB: WB:	p. 56 p. 49	Subtract within 1,000 without regrouping by using an algorithm.
3	Addition with Regrouping Ones	p. 72	TB: WB:	p. 60 p. 53	Add two numbers up to three digits within 1,000 with regrouping in the ones place by using the algorithm.
4	Addition with Regrouping Tens	p. 75	TB: WB:	p. 64 p. 57	Add two numbers within 1,000 with regrouping in the tens place using the algorithm.
5	Addition with Regrouping Tens and Ones	p. 78	TB: WB:	p. 68 p. 61	Add two numbers within 1,000 with regrouping tens and ones.
6	Practice A	p. 81	TB: WB:	p. 72 p. 65	Practice addition within 1,000. Practice subtraction without regrouping.
7	Subtraction with Regrouping from Tens	p. 84	TB: WB:	p. 74 p. 67	Subtract two numbers within 1,000 with regrouping from the tens place.
8	Subtraction with Regrouping from Hundreds	p. 87	TB: WB:	p. 78 p. 71	Subtract two numbers within 1,000 with regrouping from the hundreds place.
9	Subtraction with Regrouping from Two Places	p. 90	TB: WB:	p. 82 p. 75	Subtract two numbers within 1,000 with regrouping from the hundreds place and the tens place.
10	Subtraction with Regrouping across Zeros	p. 93	TB: WB:	p. 86 p. 79	Subtract across zeros.
11	Practice B	p. 97	TB: WB:	p. 90 p. 83	Practice subtraction.
12	Practice C	p. 98	TB: WB:	p. 91 p. 85	Practice the addition and subtraction algorithms.
	Workbook Solutions	p. 100			

In **Dimensions Math® 1B**, students learned to add and subtract two-digit numbers using mental math strategies. Students did not learn the vertical format in **Dimensions Math® 1**.

In this chapter, students extend their knowledge of addition and subtraction from a horizontal format using mental math strategies, to using the algorithms with regrouping.

The addition algorithm, also known as the vertical algorithm or standard algorithm, is a step-by-step procedure that begins with adding the digits in the ones place, then repeats with consecutively larger places. This algorithm is important because of its simplicity: it requires only single digit computations to compute sums of any kind, including decimals.

Students will use place-value discs to enrich their conceptual understanding of the algorithm as they gain procedural fluency. Students should use the place-value discs when first solving the **Think** problems. A general procedure for demonstrating the addition algorithm with place-value discs is given here.

66 + 57

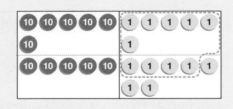

6 ones + 7 ones = 13 ones
Regroup 13 ones as 1 ten and 3 ones.
Students will write the digit 3 in the ones column and the digit 1 at the top of the tens column to represent the regrouped ten:

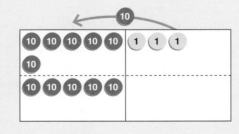

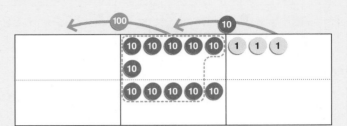

Then add the tens:
6 tens + 5 tens + 1 ten = 12 tens
Regroup 12 tens as 1 hundred and 2 tens.

As with addition, students will learn the subtraction algorithm with no regrouping first and then with regrouping:

There are not enough ones to subtract 7 ones.
Regroup 1 ten as 10 ones.

There are now 13 ones.
13 ones − 7 ones = 6 ones

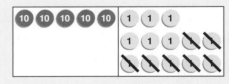

4 tens − 1 ten = 3 tens

When students are first writing equations in a vertical format, graph paper or a whiteboard with graphing lines can be helpful in aligning the digits.

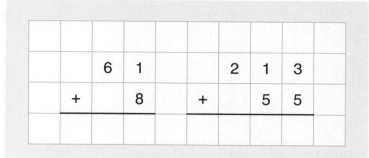

Students should understand place values to 1,000 and know their addition and subtraction facts to 20 by memory before starting this chapter.

Students may find that some problems are easily solved with a mental strategy. Encourage them to also work through the algorithm.

Students will continue to work on these algorithms in chapters on length and weight, and in Reviews. Students will build on these algorithms up to four-digit numbers in **Dimensions Math® 3A**.

When explaining the addition and subtraction processes, avoid using the terms "carrying" and "borrowing," or language like, "more on the floor, go next door," as this can lead to confusion about place value and does not help students understand the regrouping process. Ensure the terms for exchanging are used and emphasize the place being regrouped to and from.

In both the addition and subtraction algorithms, students should recognize the line is acting as an equals sign, where the terms on one side have the same value as the terms on the other side.

$$53 - 17 = 36$$

$$\begin{array}{r} 53 \\ - \ 17 \\ \hline 36 \end{array}$$

Note on managing manipulatives: Typically, in the **Dimensions Math®** series, the **Do** problems are ones that students can solve pictorially.

Due to the amounts of manipulatives used in this chapter, some suggestions on management are in order. Beginning with Lesson 3, which includes the first introduction to regrouping, consider structuring the lesson in one of these ways:

- After the **Learn** sections, provide students additional problems to work with discs and whiteboards before starting the textbook **Do** section.
- Have students work the **Do** problems using the place-value discs first, then compare them to the textbooks.
- Have students work in pairs with the place-value discs, whiteboards, and textbooks.
- When it seems students are gaining confidence and dispensing with the discs, they can copy any remaining problems from the textbook into their notebooks and work them without discs.

The goal is to ultimately have the students able to work the problems without discs.

Materials

- Graph paper
- Paper strips
- Place-value discs
- Place-value organizers

Blackline Masters

- Add 'em Up 2 numbers
- Add 'em Up Number Cards
- Add 'em Up 3 Numbers
- Add 'em Up 4 Numbers
- Greatest Difference Board
- Greatest Difference Number Cards

Activities

Fewer games and activities are included in this chapter as students will be using manipulatives extensively to learn and practice the vertical algorithms. The included activities can be used after students complete the **Do** questions, or anytime additional practice is needed. Students can also play games from the previous chapter to ensure that they have mastered their addition and subtraction facts within 20.

Objective

- Investigate addition of three-digit numbers.

Addition and Subtraction — Part 2

Mrs. Garcia wants to buy 2 paintings. How much will she pay?

Ⓐ $70
Ⓑ $140
Ⓓ $280
Ⓒ $875

51

The **Chapter Opener** is an introduction to addition with three-digit numbers.

Have students discuss which paintings Mrs. Garcia can purchase.

Students should recall strategies for adding and discuss what they think might work for adding the cost of paintings together.

Activities

▲ Race to 100

Materials: Place-value organizer, die, place-value discs

Players take turns rolling the die and placing that many place-value discs on their organizer.

When the number of discs in a column is 10 (or more), students trade them for a 10-disc.

The winner is the player who exchanges ten 10-discs for one 100-disc first.

Race to 0

Materials: Place-value organizer, die, place-value discs

Begin the game with one 100-disc and subtract the value of the die on each turn.

The winner is the first player to be out of discs.

Objective

- Add within 1,000 without regrouping by using an algorithm.

Lesson Materials

- Place-value discs
- Place-value organizer
- Graph paper
- Paper strips

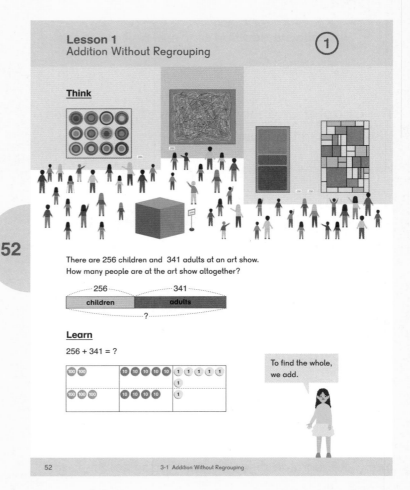

Think

Provide students with place-value discs and organizers, and allow them adequate time to work on a solution to the **Think** problem.

Ask students to relate the bar model to the word problem.

Discuss student strategies for solving the problem. Some possible student answers:

- I added just the ones together, just the tens together, and just the hundreds together.
- I made a number bond with three parts: hundreds, tens, and ones.
- I made the two numbers with the discs and counted them all up.

Learn

Have students discuss the different representations of 256 + 341. Point out to students how the discs are lined up in the textbook by place value. When adding with the algorithm, we always add like units: ones to ones, tens to tens, and hundreds to hundreds.

Using place-value discs to solve this problem, students will notice there are fewer than 10 total discs for the ones, fewer than 10 total discs for the tens, and fewer than 10 total discs for the hundreds. Because there are less than 10 total discs in each place value, students can add the digits in each place value together without any regrouping.

Teacher's Guide 2A Chapter 3   © 2017 Singapore Math Inc.

Have students work through the problem step-by-step with place-value discs.

As having discs, an organizer, and an open textbook on a desk can be overwhelming, teachers are encouraged to guide students through the steps with the discs and whiteboards first without the textbooks on their desks.

Students should record each step on a whiteboard as they are working the problem with the discs.

After working the problem, students can review the steps in the textbook.

Have students compare their solutions from **Think** with the one shown in the textbook. Ensure students use place-value language:

- 6 ones + 1 one = 7 ones
- 5 tens + 4 tens = 9 tens
- 2 hundreds + 3 hundreds = 5 hundreds

At this early stage, avoid language that doesn't ascribe place value:

- 5 + 4 is 9.
- 2 + 3 is 5.

Note that:

	2 hundreds	5 tens	6 ones	
+	3 hundreds	4 tens	1 one	
	5 hundreds	9 tens	7 ones	or 597

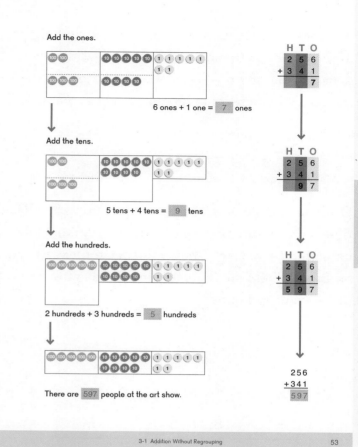

Add the ones.

6 ones + 1 one = 7 ones

Add the tens.

5 tens + 4 tens = 9 tens

Add the hundreds.

2 hundreds + 3 hundreds = 5 hundreds

There are 597 people at the art show.

3-1 Addition Without Regrouping 53

Do

As there is no regrouping, students should be able to work through these problems easily. Have them discuss what the digits represent and how they are added with place-value language. For example:

- 7 ones + 0 ones = 7 ones
- 0 tens + 9 tens = 9 tens
- 3 hundreds + 5 hundreds is 8 hundreds.

Students should line these problems up vertically. Check that they understand that the numbers line up under the correct place values. Graph paper can be used to help students line numbers in the correct place-value columns.

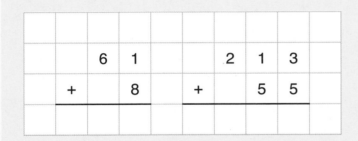

Although students may be able to solve these problems mentally, discuss the steps in the algorithm. Students can share both their mental solution and the vertical process side-by-side on a classroom board and discuss the different methods.

6 — **7** If needed, allow students to use paper strips, or draw a bar model to figure out which operation to use.

Exercise 1 · page 45

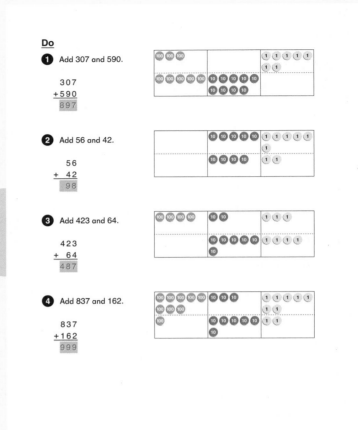

Do

1 Add 307 and 590.

$$\begin{array}{r} 307 \\ +590 \\ \hline 897 \end{array}$$

2 Add 56 and 42.

$$\begin{array}{r} 56 \\ + 42 \\ \hline 98 \end{array}$$

3 Add 423 and 64.

$$\begin{array}{r} 423 \\ + 64 \\ \hline 487 \end{array}$$

4 Add 837 and 162.

$$\begin{array}{r} 837 \\ +162 \\ \hline 999 \end{array}$$

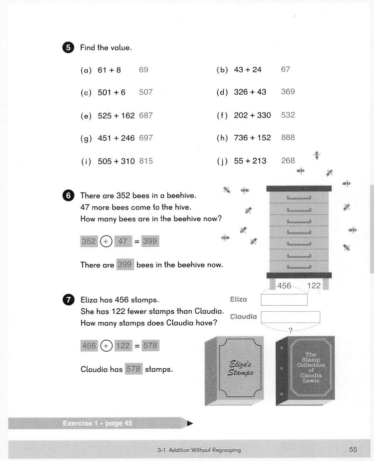

5 Find the value.

(a) 61 + 8   69

(b) 43 + 24   67

(c) 501 + 6   507

(d) 326 + 43   369

(e) 525 + 162   687

(f) 202 + 330   532

(g) 451 + 246   697

(h) 736 + 152   888

(i) 505 + 310   815

(j) 55 + 213   268

6 There are 352 bees in a beehive. 47 more bees come to the hive. How many bees are in the beehive now?

352 ⊕ 47 = 399

There are 399 bees in the beehive now.

7 Eliza has 456 stamps. She has 122 fewer stamps than Claudia. How many stamps does Claudia have?

Eliza   456   122

Claudia   ?

456 ⊕ 122 = 578

Claudia has 578 stamps.

Exercise 1 · page 45

Lesson 2 Subtraction without Regrouping

Objective

- Subtract within 1,000 without regrouping by using an algorithm.

Lesson Materials

- Place-value discs
- Place-value organizer
- Graph paper
- Paper strips

Think

Provide students with place-value discs and organizers, and allow adequate time to work on the **Think** problem.

Ask if the bar model might help them figure out whether to add or subtract.

Discuss student strategies for solving the problem. Some possible student solutions:

- I subtracted just the ones from the ones, and just the tens from the tens, and just the hundreds from the hundreds because I had enough of them.
- I made a number bond with 584 children and since one part was Cheetahs, the other part would be the Roadrunners. So I subtracted.
- I made the two numbers with the discs and subtracted them.

Learn

Have students discuss the different representations of 584 − 253. Ask students what the bar model shows. Dion reminds us that if we know the whole and one part, we can subtract to find the missing part.

Because we are subtracting fewer than the total of all discs in each place value, it is easy to simply remove the appropriate number of discs for each place.

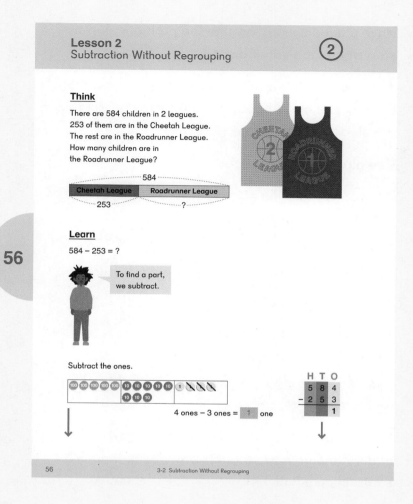

Have students work through the problem step-by-step with place-value discs.

After working the problem with discs and recording the process on whiteboards, students can discuss the steps in the textbook.

Use place-value language:

- 4 ones − 3 ones = 1 one
- 8 tens − 5 tens = 3 tens
- 5 hundreds − 2 hundreds = 3 hundreds

At this early stage, avoid language that doesn't assign to place value:

- 8 − 5 = 3
- 5 − 2 = 3

Discuss Mei's idea of how to check a subtraction problem.

Number bond language can be helpful for understanding how students can check their subtraction solutions:

A whole − part = part, or 584 − 253 = ?
and
Part + part = whole, or ? + 253 = 584

Note that:

	5 hundreds	8 tens	4 ones		
−	2 hundreds	5 tens	3 ones		
	3 hundreds	3 tens	1 one	or	331

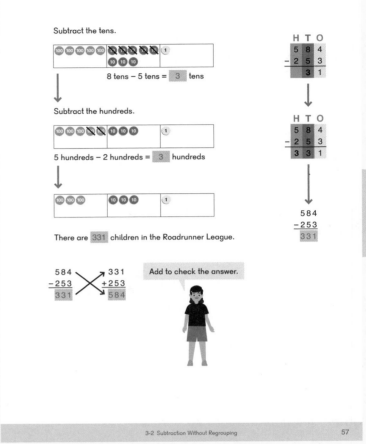

Subtract the tens.

8 tens − 5 tens = 3 tens

Subtract the hundreds.

5 hundreds − 2 hundreds = 3 hundreds

There are 331 children in the Roadrunner League.

584   331
−253   +253
331   584

Add to check the answer.

57

Teacher's Guide 2A Chapter 3

Do

Have students work through the problems with place-value discs as needed. As there is no regrouping, students should be able to work the problems without the discs.

Check that students describe the steps using place-value language.

⑤ Have students rewrite the digits vertically and ensure they are aligning the digits correctly. Graph paper can be used if needed.

Students may be able to solve these problems mentally. Students can share both their mental solution and the vertical process side-by-side on a classroom board and discuss the different methods.

⑥ Ask students how they can use Mei's strategy from **Learn** to find the missing numbers.

⑦—**⑧** If needed, allow students to use paper strips, or draw a bar model to figure out which operation to use.

Exercise 2 • page 49

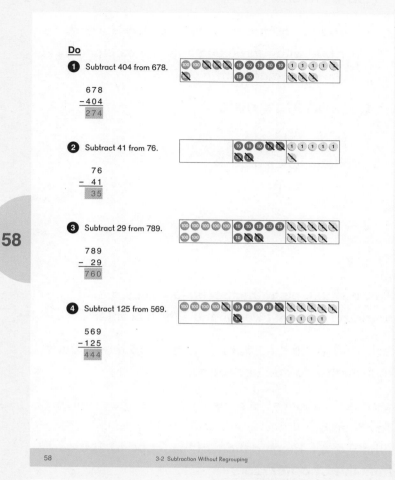

Do

① Subtract 404 from 678.

$$\begin{array}{r} 678 \\ -404 \\ \hline 274 \end{array}$$

② Subtract 41 from 76.

$$\begin{array}{r} 76 \\ -\ 41 \\ \hline 35 \end{array}$$

③ Subtract 29 from 789.

$$\begin{array}{r} 789 \\ -\ 29 \\ \hline 760 \end{array}$$

④ Subtract 125 from 569.

$$\begin{array}{r} 569 \\ -125 \\ \hline 444 \end{array}$$

58

3-2 Subtraction Without Regrouping

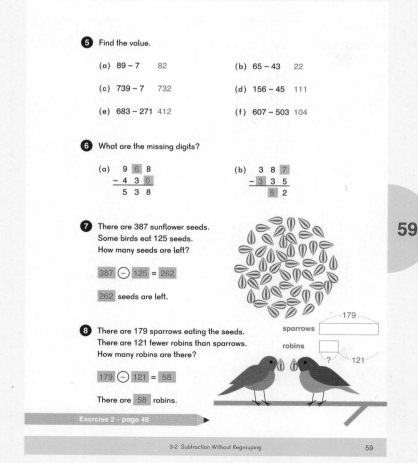

⑤ Find the value.

(a) $89 - 7$ 82

(b) $65 - 43$ 22

(c) $739 - 7$ 732

(d) $156 - 45$ 111

(e) $683 - 271$ 412

(f) $607 - 503$ 104

⑥ What are the missing digits?

(a)
$$\begin{array}{r} 9\ 6\ 8 \\ -\ 4\ 3\ 0 \\ \hline 5\ 3\ 8 \end{array}$$

(b)
$$\begin{array}{r} 3\ 8\ 7 \\ -\ 3\ 3\ 5 \\ \hline 5\ 2 \end{array}$$

⑦ There are 387 sunflower seeds. Some birds eat 125 seeds. How many seeds are left?

$387 \ominus 125 = 262$

262 seeds are left.

⑧ There are 179 sparrows eating the seeds. There are 121 fewer robins than sparrows. How many robins are there?

sparrows $\boxed{179}$

robins $\boxed{}$? ····· 121

$179 \ominus 121 = 58$

There are 58 robins.

Exercise 2 • page 49

3-2 Subtraction Without Regrouping   59

59

Lesson 3 Addition with Regrouping Ones

Objective

- Add two numbers up to three digits within 1,000 with regrouping in the ones place by using the algorithm.

Lesson Materials

- Place-value discs
- Place-value organizer
- Paper strips

Think

Provide students with place-value discs and an organizer to work through the **Think** problem.

Ask students if the bar model can help them figure out whether to add or subtract.

Discuss student strategies for solving the problem. Ask them what they can do when they have more than 9 ones. Remind them of the **Race to 100** game from the **Chapter Opener**.

Students could use the mental strategy of making 10 or another strategy to solve this problem:

$$158 + 136 = 160 + 134$$

with 2 and 134

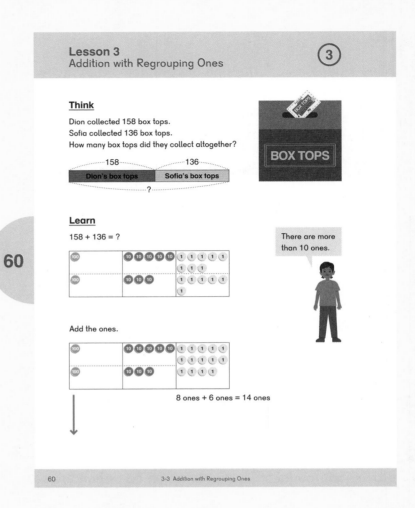

Learn

Work through the **Think** problem with the students as demonstrated in the **Learn**. Students should have the discs and organizers on their desks. Have students work along as the steps are being modeled.

Students do not need to write the numerals of the vertical algorithm until after they have worked the problem with the discs at least once.

After the students have completed the problem with discs, have them compare the methods they used in **Think** with the method shown in the textbook.

Use the following language as guidance:

- 8 ones + 6 ones = 14 ones. 14 ones is also 1 ten and 4 ones. I can regroup the ten ones as 1 ten and put the additional 10-disc in the tens column.
- Now I have 4 ones.

The textbook shows the words: 1 ten + 5 tens + 3 tens, however, the chart shows a row of 5 tens and a row of 4 tens. Discuss with the students why there is an additional ten and where it is recorded in the equation:

- 1 ten + 5 tens + 3 tens = 9 tens
- 1 hundred + 1 hundred = 2 hundreds
- 158 + 136 = 2 hundreds + 9 tens + 4 ones

Or:

	1 hundred	5 tens	8 ones		
+	1 hundred	3 tens	6 ones		
	2 hundreds	8 tens	14 ones	or	294

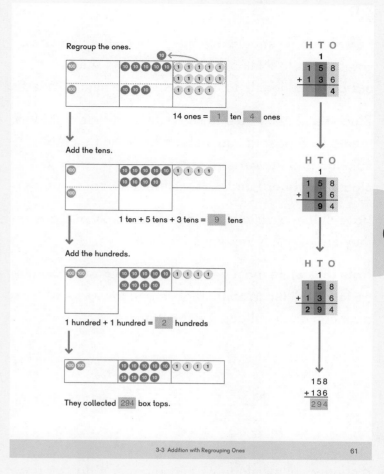

Regroup the ones.

14 ones = 1 ten 4 ones

Add the tens.

1 ten + 5 tens + 3 tens = 9 tens

Add the hundreds.

1 hundred + 1 hundred = 2 hundreds

They collected 294 box tops.

3-3 Addition with Regrouping Ones   61

Questions to ask students:

- What do we do when we have more than 9 discs in a column?
- In which column do we put the regrouped ones disc?
- Where do we show regrouping on the equation?
- What does the 1 above the 5 in the tens column stand for? Where did you see this 1 in the place-value discs?

Do

❶—❹ Students should use the pictures of the place-value discs to think about the problems. Allow struggling students to use place-value discs if needed.

❺ Students may be able to solve the first four problems mentally. Students could share both their mental solution and the vertical process side-by-side on a classroom board and discuss the different methods.

Have students rewrite the digits vertically and ensure they are aligning the digits correctly.

Note that while most people stack the greater number on top, addition is commutative and the order of addends does not change the sum:

$$
\begin{array}{r}
78 \\
+\ 302 \\
\hline
380
\end{array}
\qquad
\begin{array}{r}
302 \\
+\ \ \ 78 \\
\hline
380
\end{array}
$$

❼ If needed, allow students to use paper strips, or draw a bar model to figure out which operations to use.

Exercise 3 • page 53

Do

❶ Add 685 and 207.

$$
\begin{array}{r}
685 \\
+207 \\
\hline
892
\end{array}
$$

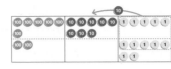

❷ Add 49 and 326.

$$
\begin{array}{r}
326 \\
+\ 49 \\
\hline
375
\end{array}
$$

❸ Add 65 and 25.

$$
\begin{array}{r}
65 \\
+\ 25 \\
\hline
90
\end{array}
$$

❹ Add 408 and 507.

$$
\begin{array}{r}
408 \\
+507 \\
\hline
915
\end{array}
$$

❺ Find the value.

(a) 45 + 7   52

(b) 38 + 48   86

(c) 609 + 8   617

(d) 78 + 302   380

(e) 175 + 209   384

(f) 746 + 234   980

❻ Explain the errors made in the calculations below and find the correct answer.

(a)
$$
\begin{array}{r}
43 \\
+\ 38 \\
\hline
711 \\
81
\end{array}
$$
✗ Wrote down total ones rather than regrouping

(b)
$$
\begin{array}{r}
43 \\
+\ 38 \\
\hline
71 \\
81
\end{array}
$$
✗ Did not add in the regrouped ten

❼ 2 years ago, there were 319 members in a club. Last year, 109 new members joined. This year, 65 new members have joined so far.

(a) At the end of last year, how many members did the club have?

319 ⊕ 109 = 428

The club had 428 members at the end of last year.

(b) How many members does the club have now?

428 ⊕ 65 = 493

The club now has 493 members.

Exercise 3 • page 53

Lesson 4 Addition with Regrouping Tens

Objective

- Add two numbers within 1,000 with regrouping in the tens place using the algorithm.

Lesson Materials

- Place-value discs
- Place-value organizer

Think

Provide students place-value discs and a place-value organizer to work through the **Think** problem.

Ask students:

- How is this problem different from the ones you solved in the previous lesson? (We have to regroup the tens.)
- How is it the same? (We can still add the places together.)

Guide students to see that the digits in the tens place, when added, will be more than 9 tens.

Discuss student strategies for solving the problem. Ask them what they can do when they have more than 9 tens.

It is unlikely that students will use a mental strategy to solve this problem.

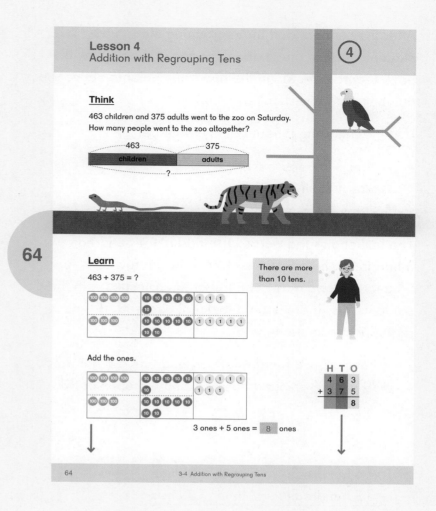

Learn

Work through the **Think** problem with the students as demonstrated in **Learn**. Students should have the discs and organizers on their desks. Have students work along as the steps are being modeled.

Students do not need to write the numerals of the vertical algorithm until after they have worked the problem with the discs at least once.

After the students have completed the problem with discs, have them compare the methods they used in **Think** with the method shown in the textbook.

After they understand the regrouping step, have students look at the textbook pages and note how the book records the "1" in the hundreds column above the numbers in the vertical equation.

Use the following language as guidance:

- 3 ones + 5 ones = 8 ones
- 6 tens + 7 tens = 13 tens
- 13 tens = 1 hundred and 3 tens

The text shows the words: 1 hundred + 4 hundreds + 3 hundreds, however, the chart shows a row of 5 hundreds and a row of 3 hundreds. Discuss with the students why there is an additional hundred. (It was regrouped from the tens.)

- 1 hundred + 4 hundreds + 3 hundreds = 8 hundreds
- 463 + 375 = 8 hundreds + 3 tens + 8 ones or 838

Or:

	4 hundreds	6 tens	3 ones	
+	3 hundreds	7 tens	5 ones	
	7 hundreds	13 tens	8 ones	or 838

Questions to ask students:

- What do we do when we have more than 9 discs in a column?
- In which column do we put the regrouped hundred disc?
- Where do we show regrouping on the equation?
- What does the 1 above the 4 in the hundreds column stand for? Where did you see this 1 in the place-value discs?

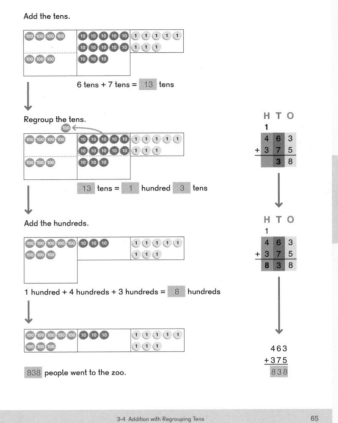

3-4 Addition with Regrouping Tens   65

Do

❶—❹ Students should use the pictures of the place-value discs to think about the problems. Allow struggling students to use place-value discs if needed.

❺ Students should rewrite the problems vertically. Ensure that they are lining up the digits correctly.

❻ Part (a) of this problem is straightforward, while (b) is more complicated as students can't just subtract to find the missing part in the tens place. The whole is actually 10 tens, not 0 tens.

Exercise 4 • page 57

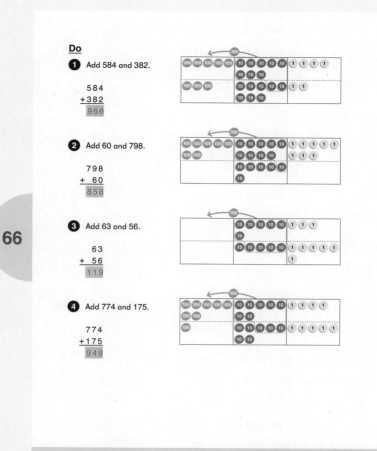

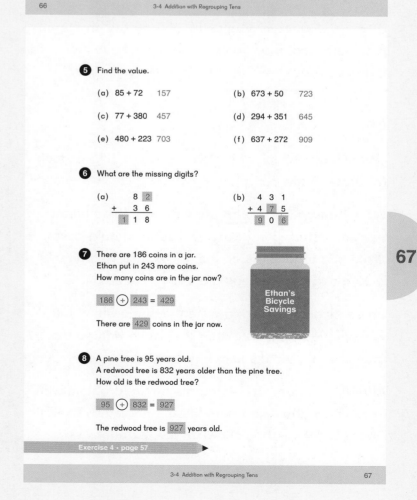

Lesson 5 Addition with Regrouping Tens and Ones

Objective

- Add two numbers within 1,000 with regrouping tens and ones.

Lesson Materials

- Place-value discs
- Place-value organizers

Think

Provide students place-value discs and organizers to work through the **Think** problem.

Students should consider what they already know about regrouping in one place. Ask students:

- How is this problem different from the ones you solved in the previous lesson? (We have to regroup more than once.)
- How is it the same? (We can still add the places together.)

Discuss student strategies for solving the problem. Ask them what they can do when they have more than 9 discs in both the ones and the tens columns.

Learn

Have students look at the bar model in the textbook (or draw on the board).

Work through the **Think** problem with the students as demonstrated in **Learn**. Students should have the discs and organizers on their desks. Have students work along as the steps are being modeled.

Students do not need to write the numerals of the vertical algorithm until after they have worked the problem with the discs at least once.

After the students have completed the problem with discs, have them compare the methods they used in **Think** to the method shown in the textbook.

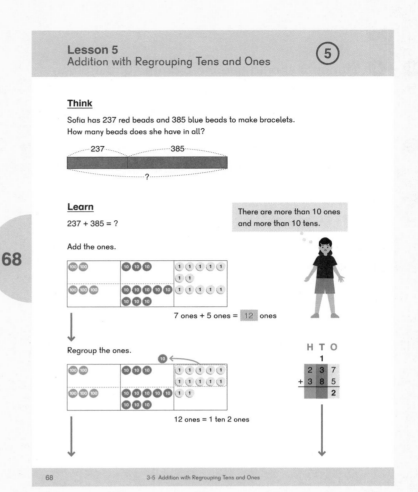

Teacher's Guide 2A Chapter 3 © 2017 Singapore Math Inc.

Use the following language as guidance:

- 7 ones plus 5 ones is 12 ones.
- 12 ones is 1 ten and 2 ones.

Ask students where we show the regrouped ten in our algorithm. (Above the tens.)

The text shows the words: 1 ten + 3 tens + 8 tens and the students should have 12 tens in their tens column.

- 12 tens is equal to 1 hundred and 2 tens.

Ask students where we show the regrouped hundred in our algorithm. (Above the hundreds.)

- 1 hundred + 2 hundreds + 3 hundreds is 6 hundreds.
- 237 + 385 = 6 hundreds + 2 tens + 2 ones
- 237 + 385 = 622

Or:

	2 hundreds	3 tens	7 ones	
+	3 hundreds	8 tens	5 ones	
	5 hundreds	11 tens	12 ones	or 622

Questions to ask students:

- What do we do when we have more than 9 discs in a column?
- In which column do we put the regrouped 1-disc? 10-disc?
- Where do we show regrouping on the equation?
- What do the 1s above the numbers stand for? Where did you see these 1s in the place-value discs?

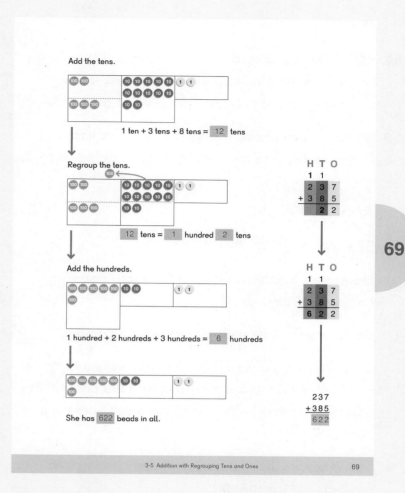

Add the tens.

1 ten + 3 tens + 8 tens = 12 tens

Regroup the tens.

12 tens = 1 hundred 2 tens

Add the hundreds.

1 hundred + 2 hundreds + 3 hundreds = 6 hundreds

She has 622 beads in all.

```
  H T O
  1 1
  2 3 7
+ 3 8 5
      2 2
```

```
  H T O
  1 1
  2 3 7
+ 3 8 5
  6 2 2
```

```
  2 3 7
+ 3 8 5
  6 2 2
```

69

Do

1—**4** Students should use the pictures of the place-value discs to think about the problems. Allow struggling students to use place-value discs if needed.

5 Students should rewrite the problems vertically. Check to ensure they are aligning the digits correctly.

Activity

▲ **501 Up**

Materials: Number Cards (BLM) 1 to 9 or playing cards, recording sheet

Each player starts with 100 as a start number.

On each turn, players draw 2 cards and make the least two-digit number they can with the cards. They add that number to their start number to create a new start number.

The winner is the first player whose running total exceeds the number 501.

Example play: Player 1 starts with 100 and draws:

He makes the number 39 and adds that to his starting number of 100. His new starting number on his next turn is 139.

Exercise 5 • page 61

Do

1 Add 462 and 348.

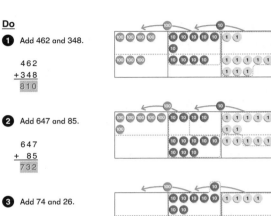

$$\begin{array}{r} 462 \\ +348 \\ \hline 810 \end{array}$$

2 Add 647 and 85.

$$\begin{array}{r} 647 \\ +\ 85 \\ \hline 732 \end{array}$$

3 Add 74 and 26.

$$\begin{array}{r} 74 \\ +\ 26 \\ \hline 100 \end{array}$$

4 Add 288 and 288.

$$\begin{array}{r} 288 \\ +288 \\ \hline 576 \end{array}$$

5 Find the value.

(a) 62 + 49   111

(b) 85 + 119   204

(c) 365 + 287   652

(d) 544 + 276   820

(e) 457 + 348   805

(f) 399 + 399   798

(g) 733 + 167   900

(h) 285 + 159   444

6 John has $259.
Samuel has $192 more than John.

(a) How much money does Samuel have?

259 (+) 192 = 451

Samuel has $451.

John | 259 | 192
Samuel | | ?

(b) How much money do they have altogether?

259 (+) 451 = 710

They have $710 altogether.

7 Catalina has $100.
She wants to buy shoes for $65 and a shirt for $37. 65 + 37 = 102
Does she have enough money? No
If not, how much more money does she need? She needs $2 more.

Exercise 5 • page 61

Lesson 6 Practice A

Objectives

- Practice addition within 1,000.
- Practice subtraction without regrouping.

Practice

After students complete the **Practice** in the textbook, have them continue adding numbers with regrouping by playing games from this or previous chapters.

Observe students to assess their mastery of the skills and provide help in small groups or to individual students as needed.

Activities

▲ Add 'em Up! 2 Numbers

Materials: Number Cards (BLM) or playing cards 1 to 9, 10-sided die, Add 'em Up 2 Numbers (BLM)

Deal 7 cards to each player. Each player uses 6 of her cards to make 2 three-digit numbers with the lowest possible sum. The extra card is a discard.

The winner is the player with the lowest sum in each round.

★ Add an extra challenge to this activity by having students roll a 10-sided die and putting the numbers they roll in the squares. Note that students must have a final three-digit sum.

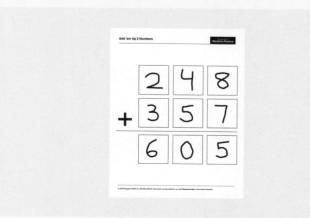

---

1 Find the value.

(a) 33 + 5   38

(b) 67 – 5   62

(c) 89 – 42   47

(d) 179 – 60   119

(e) 356 + 23   379

(f) 183 – 61   122

(g) 689 – 427   262

(h) 423 + 54   477

(i) 608 + 191   799

(j) 305 – 102   203

72

2 A farmer had 580 tomatoes.
He sold 430 of them and used the rest to make tomato sauce.
How many tomatoes did he use to make the sauce?
580 – 430 = 150
150 tomatoes

3 A photograph costs $110 and a painting costs $280.
How much more does the painting cost than the photograph?
280 – 110 = 170
$170

4 Fadiya wants to buy a coat and some shoes.
The coat costs $153 and the shoes cost $32.

(a) How much more does the coat cost than the shoes?
153 – 32 = 121; $121

(b) How much money does Fadiya need to buy both items?
153 + 32 = 185; $185

72   3-6 Practice A

▲ Race to the Sum for 2 Numbers

Materials: Add 'em Up 2 Numbers (BLM), Add 'em Up Number Cards (BLM), 2- or 3-minute timer

Cut out the numbers from Add 'em Up Number Cards (BLM) and lay them facedown in front of the players. Players draw 10 random cards each and keep them facedown in front of them.

When both players are ready, they start the timer and flip over their cards, and begin making equations with up to 9 of their cards.

Players work until the timer runs out to create a correct addition problem.

When time is up, players tally their points:

- 1 point for having a correct equation
- 2 points for the greatest sum
- 1 bonus point for using the most cards in an equation

The winner is the first player to collect 15 points.

Note: It's acceptable for the problem to use a two-digit number. The player's goal is to complete an equation by the end of the 2 minutes.

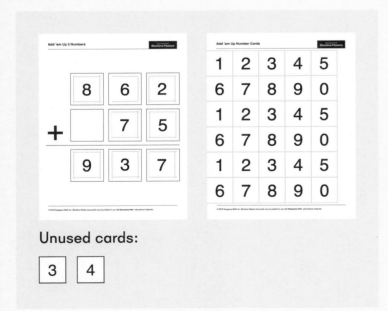

Unused cards:

3 4

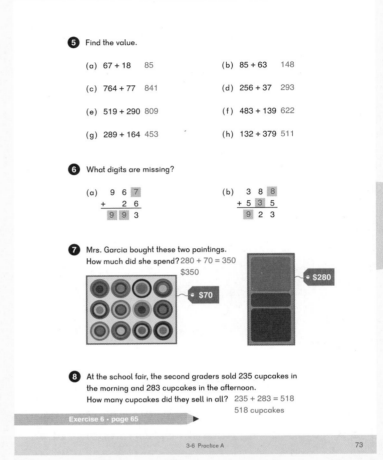

5 Find the value.

(a) 67 + 18   85
(b) 85 + 63   148
(c) 764 + 77   841
(d) 256 + 37   293
(e) 519 + 290   809
(f) 483 + 139   622
(g) 289 + 164   453
(h) 132 + 379   511

6 What digits are missing?

(a)
```
    9 6 [7]
  +   2 6
  [9] 9 3
```

(b)
```
    3 8 [8]
  + 5 [3] 5
  [9] 2 3
```

7 Mrs. Garcia bought these two paintings. How much did she spend? 280 + 70 = 350
$350

• $70   • $280

8 At the school fair, the second graders sold 235 cupcakes in the morning and 283 cupcakes in the afternoon. How many cupcakes did they sell in all?   235 + 283 = 518
518 cupcakes

Exercise 6 • page 65

3-6 Practice A   73

Activities can be extended to adding more than 2 numbers.

▲ Add 'em Up! 3 Numbers

Materials: Add 'em Up 3 Numbers (BLM), 2 modified dice with sides labeled: 1, 1, 2, 2, 3, 3 and 1 modified die with sides labeled: 1, 2, 2, 3, 4, 4

Students can play with partners or in small groups.

On each turn, one player rolls the dice. All players create and record a three-digit number based on the dice on their Add 'em Up Numbers (BLM). After 3 turns, players add up their numbers.

The winner is the player with the lowest sum in each round.

▲ Add 'em Up! 4 Numbers

Materials: Add 'em Up 4 Numbers (BLM), 1 modified die with sides labeled: 1, 1, 2, 2, 3, 3 and 1 modified die with sides labeled: 1, 2, 2, 3, 4, 4

Students can play with partners or in small groups.

On each turn, one player rolls the dice. All players create and record a two-digit number based on the dice on their Add 'em Up Numbers (BLM). After 4 turns, players add up their numbers.

The winner is the player with the greatest sum in each round.

▲ Race to the Sum for 4 Numbers

Materials: Add 'em Up 4 Numbers (BLM), 3 sets of Number Cards (BLM) or playing cards 1 to 4, 2-minute timer

Two players each draw 4 cards and make 2 two-digit numbers from the cards.

When both players have their numbers, they start the timer and copy all 4 two-digit numbers on their Add 'em Up Numbers (BLM).

Players add up the 4 numbers. When time is up, players check their answers.

- 1 point for having a correct answer
- 1 bonus point for being the only one with the correct answer

Players return the cards to the pile, shuffle the cards, and start over.

The winner is the first player to collect 6 points.

Note: Point out to students that regardless of the order of their two-digit numbers, both partners' answers should be the same.

Exercise 6 • page 65

Lesson 7 Subtraction with Regrouping from Tens

Objective

- Subtract two numbers within 1,000 with regrouping from the tens place.

Lesson Materials

- Place-value discs
- Place-value organizers

Think

Pose the **Think** problem about people at the festival and allow students time to work on a solution with place-value discs.

Students should discuss the bar model and write an equation, then find the answer.

Discuss strategies for solving the problem. Ask students what they can do when they don't have enough ones to subtract from.

Remind them of the **Race to 0** game from the **Chapter Opener**.

Learn

Discuss what Dion is thinking. The number 262 can be regrouped as 2 hundreds 5 tens and 12 ones.

Work through the **Think** problem with students as demonstrated in **Learn**. Students should have place-value discs and organizers on their desks. Have them work along as the steps are being modeled.

Students do not need to write the numerals of the vertical algorithm until after they have worked the problem with the discs at least once.

After the students have completed the problem with discs, have them compare the methods they came up with in **Think** to the method shown in the textbook.

When writing the equation in a vertical format, avoid language such as, "We always subtract the smaller number from the greater." This can cause

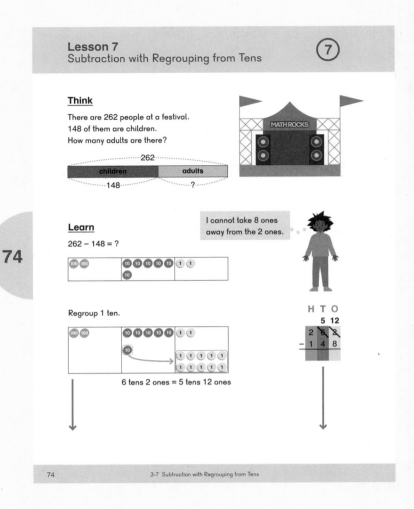

confusion, and lead students to think "8 − 2 = 6" when subtracting in the ones column because 8 is "larger" than 2. Instead, say, "There are not enough ones to subtract 8 ones."

Additionally, remind students that we have a whole and a part and are finding the other part, which they can "see" in the bar model.

Note: While algebra and higher math allow for negative numbers, it is not possible for Dion to take 8 ones discs from 2 ones discs.

Teacher's Guide 2A Chapter 3

© 2017 Singapore Math Inc.

Remind students that: 200 + 60 + 2 is the same as 200 + 50 + 12, and so:

- To keep track of the regrouping, cross off the 6 in the tens column and rewrite that as 5 tens. Cross off the 2 ones and rewrite that as 12 ones.

	2 hundreds	5 tens	12 ones
−	1 hundred	4 tens	8 ones
	1 hundred	1 tens	4 ones

- 12 ones − 8 ones = 4 ones
- 5 tens − 4 tens = 1 ten
- 2 hundreds − 1 hundred = 1 hundred
- 262 − 148 = 1 hundred + 1 ten + 4 ones, or 114

Regrouping makes subtraction much easier.

Ask students how they can use Mei's strategy of checking subtraction with an addition equation.

Questions to ask students:

- Why are we starting with the ones column?
- What do we do when we do not have enough discs in a column to subtract from?
- Why isn't the answer in the ones column 6?
- Where did the 12 written above the hundreds place come from? What is its value?
- How can we check our work?

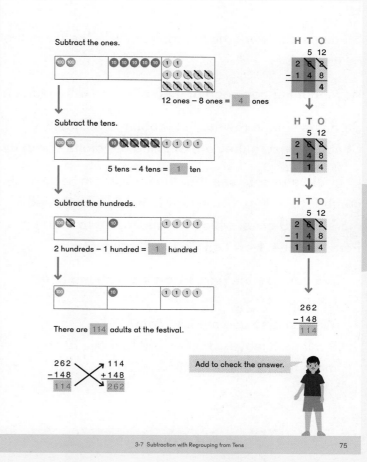

Do

Students can work these problems with the discs first as needed.

Have them check each of their answers with addition.

5 Students should rewrite the problems vertically. Check to ensure they are aligning the digits correctly.

6 (a) Help students see that the largest single-digit number (9) minus 6 can only leave 3, so we need more than 9 ones, which means having to regroup from the tens. How can we find that number?

- We can use the part + part = whole method: 6 + 5 = 11.
- We can also use mental math: 9 minus 6 leaves 3. We need to leave 5, so 2 more than 9 will leave 5: 9 + 2 = 11.

Exercise 7 • page 67

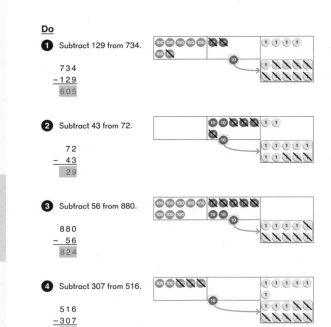

Do

1 Subtract 129 from 734.

```
  734
 −129
  605
```

2 Subtract 43 from 72.

```
  72
 − 43
  29
```

3 Subtract 56 from 880.

```
  880
 − 56
  824
```

4 Subtract 307 from 516.

```
  516
 −307
  209
```

5 Find the value.

(a) 31 − 8   23   (b) 53 − 35   18

(c) 40 − 17   23   (d) 282 − 7   275

(e) 370 − 67   303   (f) 561 − 29   532

(g) 410 − 308 102   (h) 772 − 234 538

(i) 390 − 143 247   (j) 920 − 413 507

6 What are the missing digits?

(a)
```
    8 1
  −   4 6
    3 5
```

(b)
```
    2 3 2
  + 2 7 5
    5 0 7
```

7 Tomas wants to buy a camera that costs $560.
He has $423.
How much more money does he need?

• $560

560 ⊖ 423 = 137

He needs $ 137 more.

Exercise 7 • page 67

Lesson 8 Subtraction with Regrouping from Hundreds

Objective

- Subtract two numbers within 1,000 with regrouping from the hundreds place.

Lesson Materials

- Place-value discs
- Place-value organizers

Think

Provide students with place-value discs and organizers, and pose the **Think** problem.

Ask students:

- How is this problem different from the ones you solved in the previous lesson? (We need to regroup from the hundreds.)
- How is it the same? (We are still subtracting by place.)

Discuss student strategies for solving the problem. Ask them what they can do when they don't have enough tens.

Learn

Discuss what Alex is thinking. Ask students if they have enough ones to subtract.

Work through the **Think** problem with students as demonstrated in **Learn**. Students should have the discs and organizers on their desks. Have them work along as the steps are being modeled.

Students do not need to write the numerals of the vertical algorithm until after they have worked the problem with the discs at least once.

After the students have completed the problem with discs, have them compare the methods they used in **Think** to the method shown in the textbook.

Ask students how this is similar to regrouping from the tens to the ones.

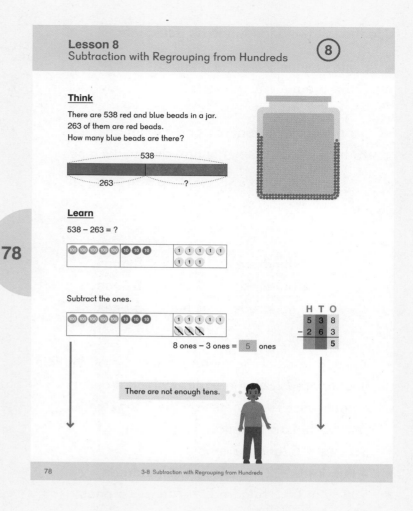

Use the following language as guidance:

- 8 ones minus 3 ones is 5 ones.
- There are not enough tens in 538 to subtract 6 tens.
- I can regroup 1 hundred and exchange it for 10 tens. Now I have 4 hundreds and 13 tens.

Remind students that 500 + 30 + 8 = 400 + 130 + 8 and so:

4 hundreds	13 tens	8 ones
− 2 hundreds	6 tens	3 ones
2 hundreds	7 tens	5 ones

- To keep track of the regrouping, cross off the 5 in the hundreds column and rewrite that as 4 hundreds. Then, cross off the 3 tens and rewrite that as 13 tens.
- 13 tens − 6 tens = 7 tens
- 4 hundreds − 2 hundreds = 2 hundreds
- 538 − 263 = 2 hundreds + 7 tens + 5 ones
- 538 − 263 = 275

Questions to ask students:

- What do we do when we do not have enough discs in a column to subtract from?
- How do we record the regrouped hundred?
- How can we check our work?

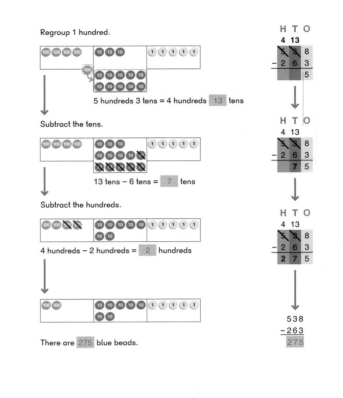

Regroup 1 hundred.

5 hundreds 3 tens = 4 hundreds 13 tens

Subtract the tens.

13 tens − 6 tens = 7 tens

Subtract the hundreds.

4 hundreds − 2 hundreds = 2 hundreds

There are 275 blue beads.

Do

Students can work these problems with the discs first as needed. Have them check each of their answers with addition.

5 Students should rewrite the problems vertically. Check to ensure they are aligning the digits correctly.

7 Students can use a bar model to help solve the problems.

Exercise 8 · page 71

80

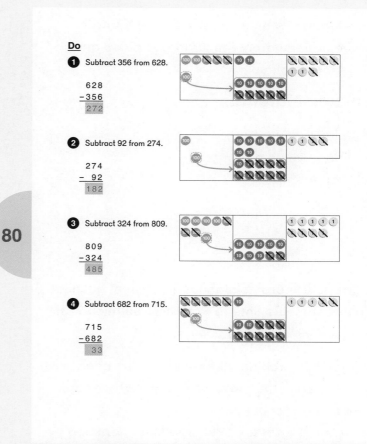

1 Subtract 356 from 628.

628
−356
272

2 Subtract 92 from 274.

274
− 92
182

3 Subtract 324 from 809.

809
−324
485

4 Subtract 682 from 715.

715
−682
33

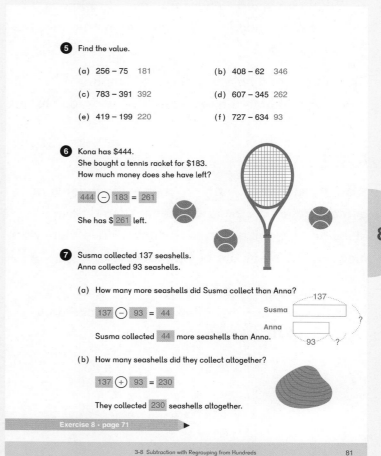

5 Find the value.

(a) 256 − 75   181

(b) 408 − 62   346

(c) 783 − 391   392

(d) 607 − 345   262

(e) 419 − 199   220

(f) 727 − 634   93

6 Kona has $444.
She bought a tennis racket for $183.
How much money does she have left?

444 ⊖ 183 = 261

She has $ 261 left.

81

7 Susma collected 137 seashells.
Anna collected 93 seashells.

(a) How many more seashells did Susma collect than Anna?

137 ⊖ 93 = 44

Susma collected 44 more seashells than Anna.

Susma
Anna
137
93
?

(b) How many seashells did they collect altogether?

137 ⊕ 93 = 230

They collected 230 seashells altogether.

Exercise 8 · page 71

Lesson 9 Subtraction with Regrouping from Two Places

Objective

- Subtract two numbers within 1,000 with regrouping from the hundreds place and the tens place.

Lesson Materials

- Place-value discs
- Place-value organizers
- Paper strips

Think

Pose the **Think** problem about the animal hospital and allow students time to work on a solution with place-value discs. Ask students:

- How is this problem different from the ones you solved in the previous lesson? (There are multiple regroupings.)
- How is it the same? (We are still subtracting by place.)
- What can you do when you have to regroup more than once? (Note the regrouping above the correct place.)

Students should note that this problem has regrouping in both the tens and the hundreds place.

Learn

Work through the **Think** problem with students as demonstrated in **Learn**. Students should have the discs and organizers on their desks. Have them work along as the steps are being modeled.

Students do not need to write the numerals of the vertical algorithm until after they have worked the problem with the discs at least once.

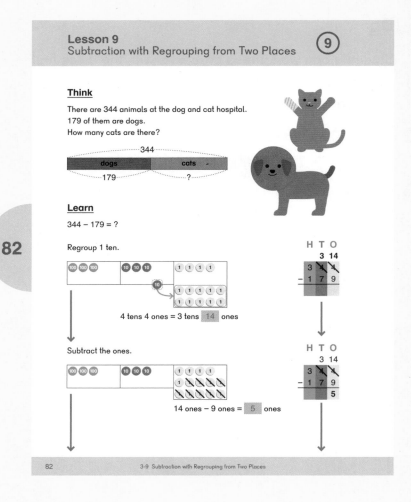

After the students have completed the problem with discs, have them compare the methods they came up with in **Think** to the method shown in the textbook.

Use the following language as guidance:

- I do not have enough ones to subtract 9 ones.
- I can regroup 1 ten and exchange it for 10 ones. Now I have 3 tens and 14 ones. (44 = 30 and 14)

To keep track of the regrouping, cross off the 4 in the tens column and rewrite that as 3 tens. Then, cross off the 4 ones and rewrite that as 14 ones.

- 14 ones − 9 ones = 5 ones

Tell students:

- Now I do not have enough tens to subtract 7 tens.
- I can regroup 1 hundred and exchange it for 10 tens. Now I have 2 hundreds and 13 tens.

To keep track of the regrouping, cross off the 3 in the hundreds column and rewrite that as 2 hundreds. Cross off the 3 tens and rewrite that as 13 tens.

- 13 tens − 7 tens = 6 tens
- 2 hundreds − 1 hundred = 1 hundred
- 344 − 179 = 1 hundred + 6 tens + 5 ones
- 344 − 179 = 165
- There are 165 cats at the animal hospital.

Or:

	hundreds	tens	ones
	2 hundreds	13 tens	14 ones
−	1 hundred	7 tens	9 ones
	1 hundred	6 tens	5 ones or 165

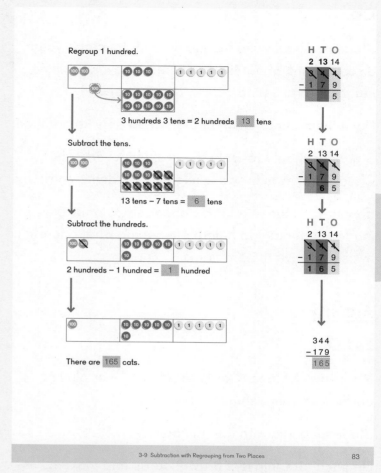

Regroup 1 hundred.

3 hundreds 3 tens = 2 hundreds 13 tens

Subtract the tens.

13 tens − 7 tens = 6 tens

Subtract the hundreds.

2 hundreds − 1 hundred = 1 hundred

There are 165 cats.

$$\begin{array}{r} 344 \\ -179 \\ \hline 165 \end{array}$$

3-9 Subtraction with Regrouping from Two Places   83

Questions to ask students:

- What do we do when we do not have enough discs in a column to subtract from?
- How do we record the regrouped discs?
- How can we check our work?

Do

Students should use the pictures of the place-value discs to think about the problems. Allow struggling students to use place-value discs if needed.

Have them check each of their answers with addition.

5 Students should re-write the problems vertically. Check to ensure they are aligning the digits correctly.

6 Encourage students to use paper strips or draw models if needed to help them decide what operation to use to solve the problems.

Activity

▲ **501 Out**

Materials: Number Cards (BLM) 1 to 9 or playing cards, recording sheet

Students begin with the number 501.

On each turn, players draw 2 cards and make a two-digit number. They subtract that number from their start number to create a new start number.

The winner is the first player whose running total gets to 0.

Example play: Player 1 starts with 501 and draws:

She makes the number 93 and subtracts that from her starting number of 501. Her new starting number on her next turn is 408.

Exercise 9 · page 75 ▶

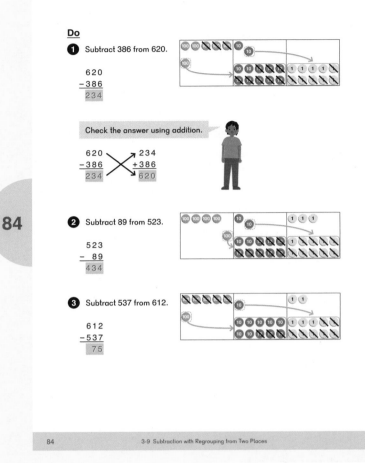

Do

1 Subtract 386 from 620.

```
  620
- 386
  234
```

Check the answer using addition.

```
  620  →  234
- 386  ╳ + 386
  234  →  620
```

2 Subtract 89 from 523.

```
  523
-  89
  434
```

3 Subtract 537 from 612.

```
  612
- 537
   75
```

84   3-9 Subtraction with Regrouping from Two Places

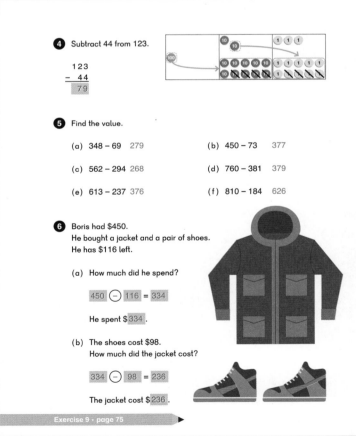

4 Subtract 44 from 123.

```
  123
-  44
   79
```

5 Find the value.

(a) 348 − 69   279
(b) 450 − 73   377
(c) 562 − 294   268
(d) 760 − 381   379
(e) 613 − 237   376
(f) 810 − 184   626

6 Boris had $450.
He bought a jacket and a pair of shoes.
He has $116 left.

(a) How much did he spend?

450 ⊖ 116 = 334

He spent $334.

(b) The shoes cost $98.
How much did the jacket cost?

334 ⊖ 98 = 236

The jacket cost $236.

Exercise 9 · page 75 ▶

3-9 Subtraction with Regrouping from Two Places   85

Lesson 10 Subtraction with Regrouping across Zeros

Objective

- Subtract across zeros.

Lesson Materials

- Place-value discs
- Place-value organizers

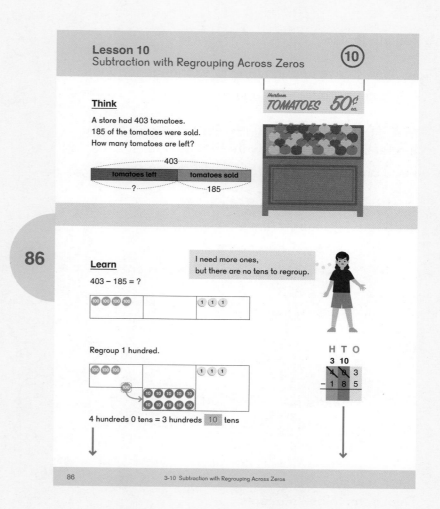

Think

Pose the **Think** problem about the tomatoes and allow student sufficient time to work on the problem. Students often have trouble subtracting across zeros, however, with the discs, it is easier to see that there is nothing to regroup from in the tens column. Ask students:

- How is this problem different from the ones you solved in the previous lesson? (There are no tens to regroup.)
- How is it the same? (We are still subtracting by place.)
- What can you do when you have to regroup more than once? (Note the regrouping above the correct place.)

Students should note that there are no tens to regroup from. They will need to regroup from the hundreds to the tens first, then from the tens to the ones.

Discuss student strategies for solving the problem.

Learn

Discuss Mei's thoughts. Work through the **Think** problem with students as demonstrated in **Learn**. Students should have the discs and organizers on their desks. Have them work along as the steps are being modeled.

Students do not need to write the numerals of the vertical algorithm until after they have worked the problem with the discs at least once.

After the students have completed the problem with discs, have them compare the methods they came up with in **Think** to the method shown in the textbook.

Use the following language as guidance:

- I do not have enough ones to subtract 5 ones.
- I can't regroup one ten and exchange it for 10 ones because I don't have any tens.
- I can exchange one hundred for 10 tens. Then I'll have some tens.

To keep track of the regrouping, cross off the 4 in the hundreds column and rewrite that as 3 hundreds. Cross off the 0 tens and rewrite that as 10 tens.

Tell students:

- Now I have enough tens to exchange for ones.
- I can regroup 1 ten and exchange it for 10 ones.

To keep track of the regrouping, I cross off the 10 in the tens column and rewrite that as 9 tens. Cross off the 3 ones and rewrite that as 13 ones.

- 13 ones − 5 ones is 8 ones.
- 9 tens − 8 tens = 1 ten
- 3 hundreds − 1 hundred = 2 hundreds
- 403 − 185 = 2 hundreds + 1 ten + 8 ones
- 403 − 185 = 218
- There are 218 tomatoes left.

Or:

	3 hundreds	9 tens	13 ones		
−	1 hundred	8 tens	5 ones		
	2 hundreds	1 ten	8 ones	or	218

Questions to ask students:

- What do we do when we do not have enough discs in a column to subtract from?
- How do we record the regrouped discs?
- Why is there a 10 and a 9 above the tens place?
- What do we do when there's a 0 in a place?
- How can we check our work?

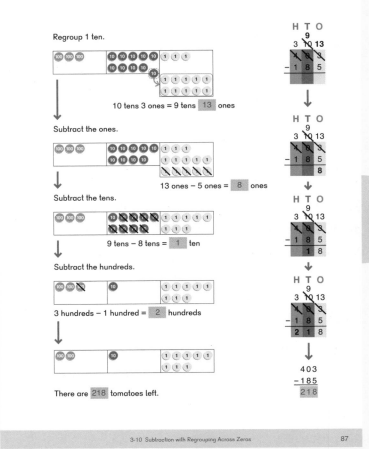

Regroup 1 ten.

10 tens 3 ones = 9 tens 13 ones

Subtract the ones.

13 ones − 5 ones = 8 ones

Subtract the tens.

9 tens − 8 tens = 1 ten

Subtract the hundreds.

3 hundreds − 1 hundred = 2 hundreds

There are 218 tomatoes left.

403
−185
218

Do

Students can work these problems with the discs first as needed. Have them check each of their answers with addition.

Focus on what Sofia and Dion are thinking and how it relates to the pictures. Ask students, "Can you follow Sofia's thoughts on the place-value discs chart?"

Students might say, "There were no tens so she exchanged 1 hundred for 10 tens, and then exchanged 1 ten for 10 ones. Now there are enough ones."

6 Students should note that when regrouping from the tens, the student forgot to record the 9 tens left. The subtraction should have been 9 tens − 8 tens = 1 ten, or the correct answer of 515.

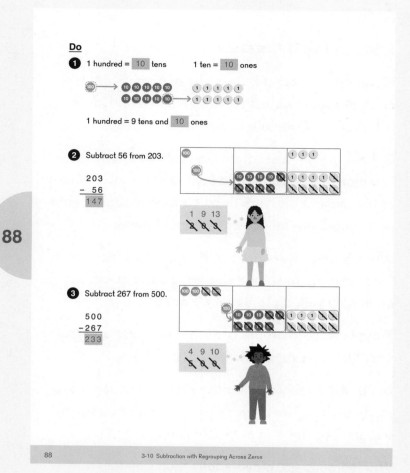

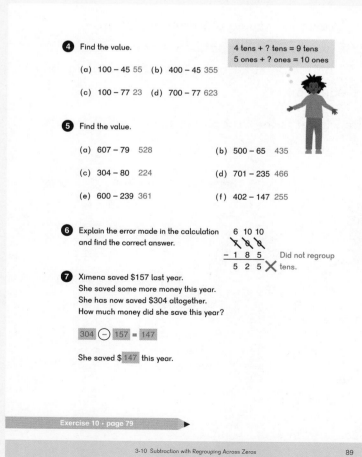

Activity

▲ Race to the Difference

Exercise 10 • page 79

Materials: Greatest Difference Boards (BLM) for each player, Greatest Difference Number Cards (BLM), 2- or 3-minute timer

Cut out the numbers from Greatest Difference Number Cards (BLM) and lay them facedown in front of the players. Players draw 10 random cards each and keep them facedown in front of them.

When both players are ready, they start the timer and flip over their cards, and begin making equations with up to 9 of their cards.

Players work to create a correct subtraction problem until the timer runs out.

Note: It's acceptable for the problem to use a two-digit number. The player's goal is to complete an equation by the end of the 2 minutes.

When time is up, players tally their points:

* 1 point for having a correct equation
* 2 points for the greatest difference
* 1 bonus point for using the most cards in their equation

The winner is the first player to collect 15 points.

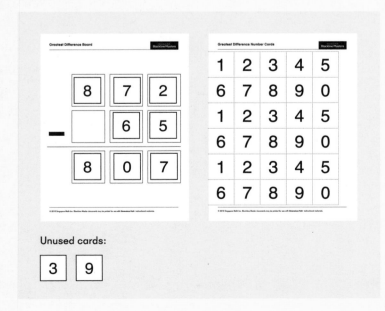

Unused cards:

Teacher's Guide 2A Chapter 3   © 2017 Singapore Math Inc.

Lesson 11 Practice B

Objective

- Practice subtraction.

Lesson Materials

- Place-value discs
- Place-value organizers

Practice

Students will continue to practice both addition and subtraction in **Practice C**.

Allow students to use place-value discs and organizers as needed.

Activity

▲ Greatest Difference

Materials: Number Cards (BLM) 1 to 9 or playing cards

Deal 7 cards to each player. Each player uses 6 of their cards to make 2 three-digit numbers with the greatest possible difference. The extra card is a discard.

Students may have a two-digit or three-digit difference. The winner is the player with the greatest difference in each round.

Exercise 11 • page 83

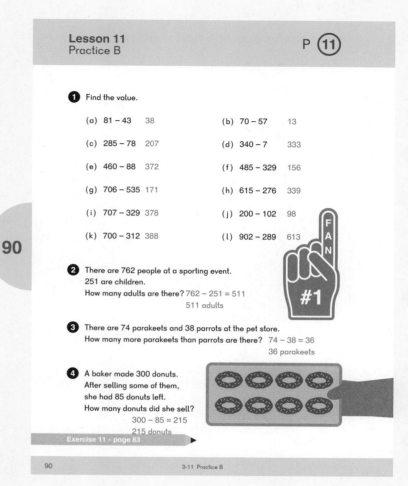

Lesson 11
Practice B
P (11)

1 Find the value.

(a) 81 – 43   38

(b) 70 – 57   13

(c) 285 – 78   207

(d) 340 – 7   333

(e) 460 – 88   372

(f) 485 – 329   156

(g) 706 – 535   171

(h) 615 – 276   339

(i) 707 – 329   378

(j) 200 – 102   98

(k) 700 – 312   388

(l) 902 – 289   613

2 There are 762 people at a sporting event. 251 are children.
How many adults are there? 762 – 251 = 511
511 adults

3 There are 74 parakeets and 38 parrots at the pet store.
How many more parakeets than parrots are there?   74 – 38 = 36
36 parakeets

4 A baker made 300 donuts.
After selling some of them, she had 85 donuts left.
How many donuts did she sell?
300 – 85 = 215
215 donuts

Exercise 11 • page 83

90   3-11 Practice B

Lesson 12 Practice C

Objective

- Practice the addition and subtraction algorithms.

2 dog

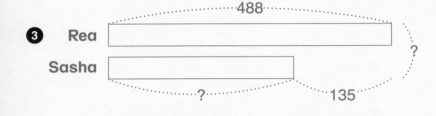

cat

3 Rea

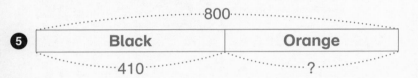

Sasha

5

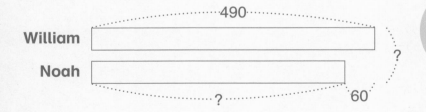

| Black | Orange |

6 This is the first time a two-step problem is given without the steps being broken into (a) and (b). Discuss the information that is known with the students.

For example, "We know how much William saved, and we know that he saved more than Noah. Can we figure out how much Noah saved and then add that to what William saved?"

William

Noah

Lesson 12
Practice C   P ⑫

1 Find the value.

(a) 397 + 8   405   (b) 789 − 2   787

(c) 403 + 69   472   (d) 89 − 22   67

(e) 603 + 91   694   (f) 276 − 30   246

(g) 213 + 224   437   (h) 375 − 12   363

(i) 856 − 182   674   (j) 207 − 139   68

2 Xavier is adopting a cat and a dog from a shelter.
It costs $307 to adopt the dog.
It costs $96 to adopt the cat.

(a) How much less does it cost to adopt the cat than the dog?
307 − 96 = 211; $211
(b) How much money does he need to adopt both pets?
307 + 96 = 403; $403

3 Rea made 488 buttons to sell.
Sasha made 135 fewer buttons than Rea.

(a) How many buttons did Sasha make?
488 − 135 = 353; 353 buttons
(b) How many buttons did they make altogether?
488 + 353 = 841; 841 buttons

91

92

4 True or false?

(a) 64 + 72 = 74 + 62   True. Digits are the same in each place.

(b) 78 − 52 = 87 − 25   False. 7 tens − 5 tens < 8 tens − 2 tens

(c) 256 + 37 < 275 + 73   True. 5 tens + 3 tens < 7 tens + 7 tens

(d) 183 − 61 < 183 + 8   True. Subtracting versus adding to same number..

(e) 100 + 54 = 200 − 45   False. 154 < 155

(f) 600 − 102 > 498 + 102   False. 600 − 102 = 498 which will be less
than adding to 498.

(g) 267 + 499 < 456 + 389   True. 766 < 845, or, sum of hundreds is greater.

(h) 914 − 186 > 257 + 176
True. 9 hundreds − 1 hundreds > 2 hundreds + 1 hundreds

5 There are 800 beads in a jar.
410 of them are black and the rest are orange.

(a) Are there more black beads or orange beads?
800 − 410 = 390; 390 orange beads, more black beads.
(b) How many more are there of one color than the other?
410 − 390 = 20; 20 more black beads

6 William saved $490.   First find how much Noah saved.
He saved $60 more than Noah.   490 − 60 = 430; $430
How much did they both save?   Then add the amounts both saved.
490 + 430 = 920; $920

Exercise 12 • page 85

Activity

▲ 3 in a Row

Materials: Hundred chart, playing cards, counters (different color for each player)

Addition: Deal players 4 cards. On each round, players make 2 two-digit numbers with a sum on the hundred chart.

Tens and face cards can be 0. If a 0 is drawn, it can be used for ones or discarded so that one of the numbers will be a one-digit number.

If it is not possible to make two numbers with a sum less than 100, the student either discards the highest digit and draws another card, or adds a two-digit to a one-digit number.

When Player 1 has found the sum, she places her counter on the sum on the hundreds board. Player 2 then places his counter. Discard those cards and deal 4 more cards.

Play continues until a player has three counters in a row, column, or diagonally.

If a number is already covered, the digits have to be rearranged to make a different sum.

Subtraction: Players find the difference instead of the sum.

Brain Works

★ Toothpick Problem

Materials: 10 toothpicks or markers

Use 10 toothpicks or markers to make the figure below. Can you move only 2 sticks and end up with only 2 squares?

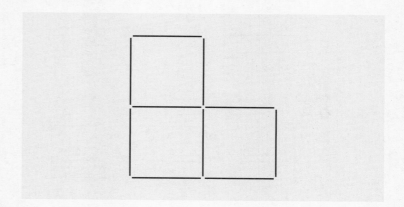

Exercise 12 • page 85

Chapter 3 Addition and Subtraction — Part 2

Exercise 1

Basics

1 (a) Add 254 and 714.

	2	5	4
+	7	1	4
	9	6	8

254 + 714 = **968**

(b) Add 32 and 836.

	8	3	6
+		3	2
	8	6	8

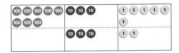

32 + 836 = **868**

Practice

2 Add.

(a)
	4	2
+	3	0
	7	2

(b)
	2	8
+	6	1
	8	9

(c)
	7	3
+	2	5
	9	8

---

3 Add.

	4	2	3
+		1	6
	4	3	9

R

	5	6	3
+		3	5
	5	9	8

D

	3	0	4
+		9	2
	3	9	6

C

	2	1	6
+	5	7	0
	7	8	6

I

	7	4	2
+	2	5	6
	9	9	8

O

	3	8	4
+	4	0	5
	7	8	9

L

	2	7	1
+	3	0	4
	5	7	5

C

	5	5	6
+	3	4	3
	8	9	9

E

	8	2	3
+	1	7	3
	9	9	6

O

What animal sometimes eats rocks?
Write the letters that match the answers above to find out.

C	R	O	C	O	D	I	L	E
575	439	998	396	996	598	786	789	899

---

4 A skateboard costs $35.
A scooter costs $23 more than the skateboard.
How much does the scooter cost?

	3	5
+	2	3
	5	8

The scooter costs $ **58** .

5 Eli has read 242 pages in a book so far.
He needs to read another 56 pages to finish the book.
How many pages does the book have?

2	4	2
+		5	6
2	9	8

The book has **298** pages.

6 After buying a guitar for $512, Makayla had $142 left.
How much money did she have at first?

5	1	2
+	1	4	2
6	5	4

She had $ **654** at first.

---

7 Complete the cross-number puzzle using the clues.

	A	B	C		D		
	4	9	7		6		
		E	6	8		F 9	5	G 8
	H 8	6	0		1		5
	4				I 7	6	J 9
	K 7	L 5		M 8		3		7
		8		N 9	7	5		4
		O 3	4	2			

Across

497	A	456 + 41
68	E	36 + 32
958	F	605 + 353
860	H	840 + 20
769	I	306 + 463
75	K	32 + 43
975	N	911 + 64
342	O	231 + 111

Down

B	104 + 862	966
C	530 + 250	780
D	141 + 550	691
G	424 + 432	856
H	122 + 725	847
I	315 + 420	735
J	453 + 521	974
L	453 + 130	583
M	720 + 172	892

Teacher's Guide 2A Chapter 3

Exercise 2

Basics

1 (a) Subtract 514 from 756.

	7	5	6
−	5	1	4
	2	4	2

756 − 514 = **242**

(b) Subtract 36 from 489.

	4	8	9
−		3	6
	4	5	3

489 − 36 = **453**

Practice

2 Subtract.

(a)
	4	2
−	3	0
	1	2

(b)
	6	8
−	3	1
	3	7

(c)
	7	5
−	2	2
	5	3

3 Subtract.

	3	7	9
−		4	2
	3	3	7

	5	8	7
−		4	5
	5	4	2

	2	9	5
−		7	3
	2	2	2

	9	7	3
−	1	7	2
	8	0	1

	9	0	8
−	3	0	3
	6	0	5

	6	9	8
−	1	3	6
	5	6	2

	9	5	7
−	4	4	1
	5	1	6

	6	8	7
−	6	7	4
		1	3

	4	9	8
−	1	8	0
	3	1	8

Color the spaces that contain the answers to find out what animal Emma's pet is.

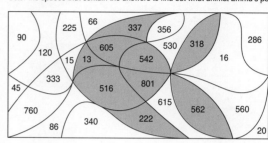

4 There are 68 golden hamsters at a pet store.
There are 25 fewer dwarf hamsters than golden hamsters.
How many dwarf hamsters are there?

		6	8
−		2	5
		4	3

There are ___43___ dwarf hamsters.

5 A book has 294 pages.
Kalama has read 62 pages.
How many more pages does she still have to read?

	2	9	4
−		6	2
	2	3	2

She has ___232___ more pages to read.

6 Laila had $398.
After buying some clothes, she had $142 left.
How much money did she spend on clothes?

	3	9	8
−	1	4	2
	2	5	6

She spent $ ___256___ on clothes.

7 Complete the cross-number puzzle using the clues.

A 9		B 5	0	C 4		
3		5		D 2	E 8	F 3
G 6	H 2		I 2		J 2	5	K 9
	3		L 3	0	6		2
	M 8	N 1	4			O 3	4
		0		P 4	1	0
	Q 1	3	3			1

Across			Down		
504	B 807 − 303		A 968 − 32	936
283	D 498 − 215		B 76 − 21	55
62	G 97 − 35		C 95 − 53	42
259	J 959 − 700		E 997 − 171	826
306	L 579 − 273		F 75 − 40	35
814	M 836 − 22		H 648 − 410	238
34	O 694 − 660		I 345 − 111	234
410	P 816 − 406		K 988 − 64	924
133	Q 843 − 710		N 489 − 388	101
			O 905 − 604	301
			P 264 − 221	43

Teacher's Guide 2A Chapter 3

Exercise 3

Basics

1. (a) Add 356 and 8.

```
    1
    3  5  6
  +       8
    3  6  4
```

356 + 8 = 364

(b) Add 356 and 28.

```
    1
    3  5  6
  +    2  8
    3  8  4
```

356 + 28 = 384

(c) Add 356 and 528.

```
    1
    3  5  6
  +  5  2  8
    8  8  4
```

356 + 528 = 884

Practice

2. Add 547 and 219.

```
    5  4  7
  +  2  1  9
    7  6  6
```

547 + 219 = 766

Students can write the lesser number on top when rewriting the expression vertically, but later, without a grid, it will be easier to align digits if the greater number is written first.

3. 808 + 3 = 811

```
    8  0  8
  +       3
    8  1  1
```

9 + 947 = 956

```
       9  4  7
  +          9
       9  5  6
```

65 + 25 = 90

```
          6  5
  +       2  5
          9  0
```

44 + 39 = 83

```
       4  4
  +    3  9
       8  3
```

856 + 29 = 885

```
    8  5  6
  +    2  9
    8  8  5
```

74 + 319 = 393

```
       3  1  9
  +       7  4
       3  9  3
```

4. Write the missing digits.

(a)
```
    9  4  9
  +    3  5
    9  8  4
```

(b)
```
    6  3  6
  +  2  5  7
    8  9  3
```

5. Add.

437 + 237
```
    4  3  7
  +  2  3  7
    6  7  4
```

128 + 566
```
    1  2  8
  +  5  6  6
    6  9  4
```

229 + 242
```
    2  2  9
  +  2  4  2
    4  7  1
```

555 + 239
```
    5  5  5
  +  2  3  9
    7  9  4
```

564 + 327
```
    5  6  4
  +  3  2  7
    8  9  1
```

608 + 378
```
    6  0  8
  +  3  7  8
    9  8  6
```

486 + 106
```
    4  8  6
  +  1  0  6
    5  9  2
```

218 + 747
```
    2  1  8
  +  7  4  7
    9  6  5
```

434 + 149
```
    4  3  4
  +  1  4  9
    5  8  3
```

Color the spaces that contain the answers to help the bird find its home.

986	694	955	964	992	659
896	674	965	794	462	573
764	582	573	592	891	664
582	976	574	991	471	583

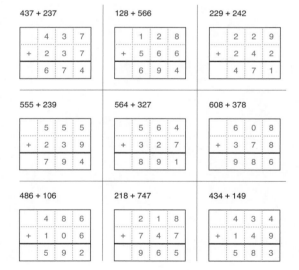

6. Megan has two plum trees.
She picked 135 plums from one tree and 157 from the other tree.
How many plums did she pick?

```
    1  3  5
  +  1  5  7
    2  9  2
```

She picked 292 plums.

7. Megan made 106 jars of jam the first week.
The second week she made 29 more jars of jam than the week before.

(a) How many jars of jam did she make the second week?

```
    1  0  6
  +    2  9
    1  3  5
```

She made 135 jars of jam the second week.

(b) How many jars of jam did she make in all both weeks?

```
    1  0  6
  +  1  3  5
    2  4  1
```

She made 241 jars of jam both weeks.

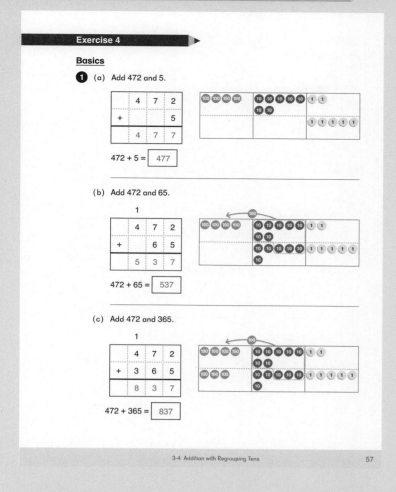

Exercise 4

Basics

1. (a) Add 472 and 5.

472 + 5 = 477

(b) Add 472 and 65.

472 + 65 = 537

(c) Add 472 and 365.

472 + 365 = 837

3-4 Addition with Regrouping Tens   57

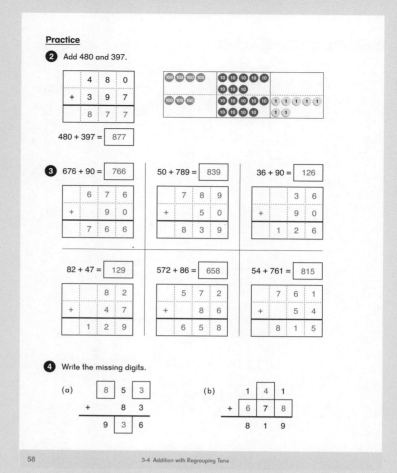

Practice

2. Add 480 and 397.

480 + 397 = 877

3. 676 + 90 = 766   50 + 789 = 839   36 + 90 = 126

82 + 47 = 129   572 + 86 = 658   54 + 761 = 815

4. Write the missing digits.

(a)   8 5 3
    +   8 3
    9 [3] 6

(b)   1 4 1
    + 6 7 8
    8 1 9

58   3-4 Addition with Regrouping Tens

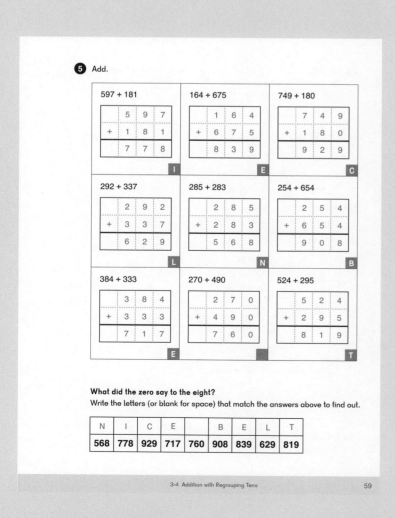

5. Add.

597 + 181 = 778   **I**
164 + 675 = 839   **E**
749 + 180 = 929   **C**

292 + 337 = 629   **L**
285 + 283 = 568   **N**
254 + 654 = 908   **B**

384 + 333 = 717   **E**
270 + 490 = 760   **S**
524 + 295 = 819   **T**

What did the zero say to the eight?
Write the letters (or blank for space) that match the answers above to find out.

N	I	C	E		B	E	L	T
568	778	929	717	760	908	839	629	819

3-4 Addition with Regrouping Tens   59

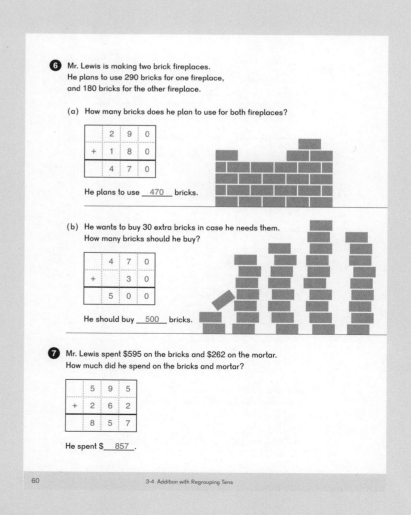

6. Mr. Lewis is making two brick fireplaces.
He plans to use 290 bricks for one fireplace,
and 180 bricks for the other fireplace.

(a) How many bricks does he plan to use for both fireplaces?

  2 9 0
+ 1 8 0
  4 7 0

He plans to use __470__ bricks.

(b) He wants to buy 30 extra bricks in case he needs them.
How many bricks should he buy?

  4 7 0
+   3 0
  5 0 0

He should buy __500__ bricks.

7. Mr. Lewis spent $595 on the bricks and $262 on the mortar.
How much did he spend on the bricks and mortar?

  5 9 5
+ 2 6 2
  8 5 7

He spent $__857__.

60   3-4 Addition with Regrouping Tens

Exercise 5

Basics

1 (a) Add 457 and 4.

```
      1
    4 5 7
  +     4
    4 6 1
```

457 + 4 = 461

(b) Add 457 and 84.

```
    1 1
    4 5 7
  +   8 4
    5 4 1
```

457 + 84 = 541

(c) Add 457 and 384.

```
    1 1
    4 5 7
  + 3 8 4
    8 4 1
```
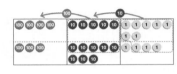

457 + 384 = 841

Practice

2 Add 376 and 394.

```
    3 7 6
  + 3 9 4
    7 7 0
```

376 + 394 = 770

3

56 + 95 = 151
```
      5 6
  +   9 5
    1 5 1
```

84 + 66 = 150
```
      8 4
  +   6 6
    1 5 0
```

49 + 59 = 108
```
      4 9
  +   5 9
    1 0 8
```

799 + 92 = 891
```
    7 9 9
  +   9 2
    8 9 1
```

683 + 18 = 701
```
    6 8 3
  +   1 8
    7 0 1
```

36 + 886 = 922
```
    8 8 6
  +   3 6
    9 2 2
```

4 Write the missing digits.

(a)
```
    1 3 9
  +   8 3
    2 2 2
```

(b)
```
    1 1 7
  + 6 8 5
    8 0 2
```

---

5 Add.

145 + 695
```
    1 4 5
  + 6 9 5
    8 4 0
```
B

258 + 278
```
    2 5 8
  + 2 7 8
    5 3 6
```
H

297 + 137
```
    2 9 7
  + 1 3 7
    4 3 4
```
I

262 + 598
```
    2 6 2
  + 5 9 8
    8 6 0
```
G

479 + 429
```
    4 7 9
  + 4 2 9
    9 0 8
```
T

388 + 269
```
    3 8 8
  + 2 6 9
    6 5 7
```
N

193 + 407
```
    1 9 3
  + 4 0 7
    6 0 0
```
E

367 + 576
```
    3 6 7
  + 5 7 6
    9 4 3
```
H

177 + 783
```
    1 7 7
  + 7 8 3
    9 6 0
```
O

What is the strongest bone in your body?
Write the letters that match the answers above to find out.

T	H	I	G	H	B	O	N	E
908	943	434	860	536	840	960	657	600

---

6 The second and third graders from Sydney's school went to the aquarium.
There were 239 second graders and 383 third graders.
How many students went on the trip that day?

```
    2 3 9
  + 3 8 3
    6 2 2
```

622 students went on the trip.

---

7 The cost for snacks at the aquarium for students was $687.
The cost for snacks for the teachers was $125.
How much did the snacks cost?

```
    6 8 7
  + 1 2 5
    8 1 2
```
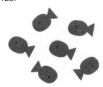

The snacks cost $812.

---

8 In one display, there were 158 fish.
In another, there were 787 fish.
How many fish were in the two displays?

```
    1 5 8
  + 7 8 7
    9 4 5
```

There were 945 fish in the two displays.

Exercise 6

Check

1 Which two numbers below will give the greatest answer when added together? Find the answer.

| 249 | 294 | 75 | 386 |

$$\boxed{386} + \boxed{294} = \boxed{680}$$

2 What digit is in the tens place for the value of each of the following:

Tens digit

(a) 241 + 537 $\boxed{7}$

(b) 135 + 438 $\boxed{7}$

(c) 183 + 447 $\boxed{3}$

(d) 466 + 327 $\boxed{9}$

While students can first do the addition, some students may be able determine the tens digit without calculating the sum completely.

3 Write >, <, or = in the ◯.

(a) 824 + 163 ◯< 456 + 541

(b) 635 + 82 ◯> 542 + 161

(c) 357 + 484 ◯= 454 + 387

(d) 836 − 412 ◯> 168 + 225

4 Add.

(a) 8 + 7 + 5 = $\boxed{20}$

(b) 70 + 90 + 60 = $\boxed{220}$

(c) 300 + 200 + 100 = $\boxed{600}$

(d)
	3	7	8
		2	9	7
+		1	6	5
		8	4	0

5 (a)
	4	9	9
		7	3
+		8	2
	6	5	4

(b)
	1	2	9
	4	3	6
+		6	8
	6	3	3

6 There are 365 men, 314 women, and 278 children in a club.

(a) How many more men than women are there?

365 − 314 = 51

There are ___51___ more men than women.

(b) How many people are there in the club altogether?

365 + 314 + 278 = 957

There are ___957___ people in the club.

Exercise 7

Basics

1 (a) Subtract 7 from 863.

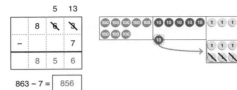

$863 - 7 = \boxed{856}$

(b) Subtract 37 from 863.

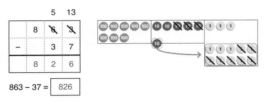

$863 - 37 = \boxed{826}$

(c) Subtract 537 from 863.

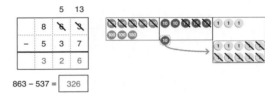

$863 - 537 = \boxed{326}$

Practice

2 Subtract 356 from 674.

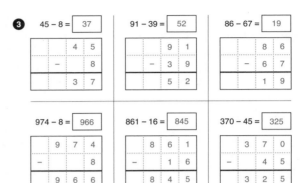

$674 - 356 = \boxed{318}$

3
$45 - 8 = \boxed{37}$ $91 - 39 = \boxed{52}$ $86 - 67 = \boxed{19}$

$974 - 8 = \boxed{966}$ $861 - 16 = \boxed{845}$ $370 - 45 = \boxed{325}$

4 Write the missing digits.

(a)

	6	8	1
−		3	5
	6	4	6

(b)

	9	8	2
−	2	0	5
	7	7	7

5 Subtract.

482 − 259

	4	8	2
−	2	5	9
	2	2	3

374 − 127

	3	7	4
−	1	2	7
	2	4	7

660 − 349

	6	6	0
−	3	4	9
	3	1	1

633 − 205

	6	3	3
−	2	0	5
	4	2	8

671 − 357

	6	7	1
−	3	5	7
	3	1	4

885 − 477

	8	8	5
−	4	7	7
	4	0	8

687 − 549

	6	8	7
−	5	4	9
	1	3	8

973 − 919

	9	7	3
−	9	1	9
		5	4

591 − 154

	5	9	1
−	1	5	4
	4	3	7

What gets bigger the more you take away?
Shade in rectangles that contain the answers.
There may be more than one for each number.

139	248	223	314		127	437	807	36	408
45	324		724	333		328	117		888
437	779	18		329		29	601		138
427	500	47		171		97	559		321
817	999		54	247	226		311	123	428

6 Jordan's science club raised $340 for a trip to the science center.
After buying tickets for $106, they donated the rest to the science center.
How much did they donate to the science center?

	3	4	0
−	1	0	6
	2	3	4

They donated $ __234__ to the science center.

7 An elk exhibit said scientists counted 873 elk in County A
and 249 elk in County B.
How many more elk did they count in County A than County B?

	8	7	3
−	2	4	9
	6	2	4

They counted __624__ more elk in County A than County B.

8 A cat exhibit said that cats have 244 bones.
Human adults have 206 bones.
How many fewer bones do human adults have than cats?

	2	4	4
−	2	0	6
		3	8

Humans have __38__ fewer bones than cats.

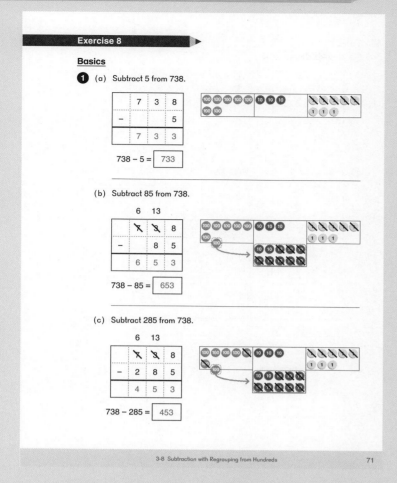

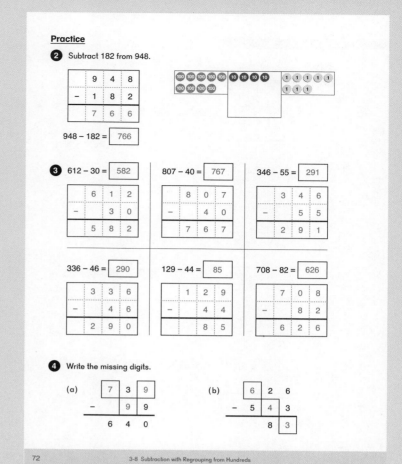

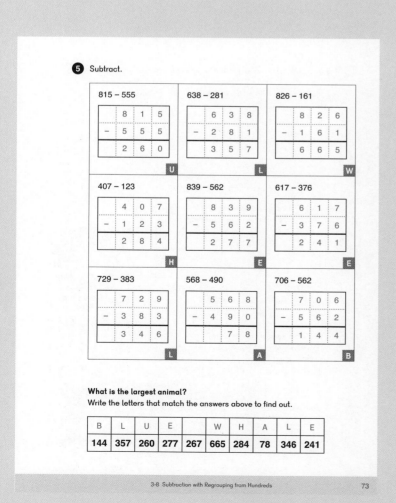

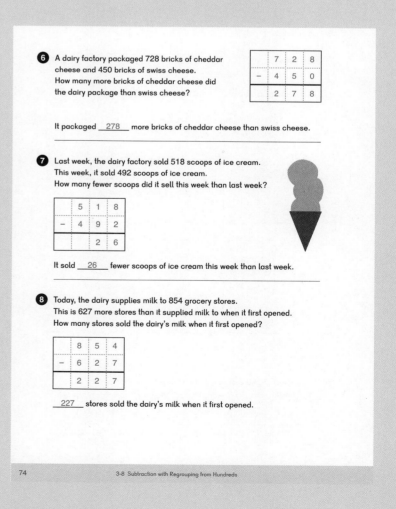

Exercise 9

Basics

1 (a) Subtract 7 from 842.

$$\begin{array}{c} \quad\quad 3 \;\; 12 \\ \begin{array}{r} 8\;\;\cancel{4}\;\;\cancel{2} \\ -\quad\quad 7 \\ \hline 8\;\;3\;\;5 \end{array} \end{array}$$

 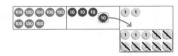

842 − 7 = 835

(b) Subtract 67 from 842.

$$\begin{array}{c} \;\; 7 \;\; 13 \;\; 12 \\ \begin{array}{r} \cancel{8}\;\;\cancel{4}\;\;\cancel{2} \\ -\;\;6\;\;7 \\ \hline 7\;\;7\;\;5 \end{array} \end{array}$$

 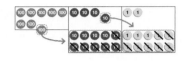

842 − 67 = 775

(c) Subtract 467 from 842.

$$\begin{array}{c} \;\; 7 \;\; 13 \;\; 12 \\ \begin{array}{r} \cancel{8}\;\;\cancel{4}\;\;\cancel{2} \\ -\;4\;\;6\;\;7 \\ \hline 3\;\;7\;\;5 \end{array} \end{array}$$

842 − 467 = 375

Practice

2 Subtract 685 from 974.

$$\begin{array}{r} 9\;\;7\;\;4 \\ -\;6\;\;8\;\;5 \\ \hline 2\;\;8\;\;9 \end{array}$$

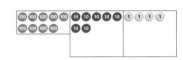 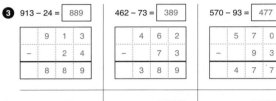

974 − 685 = 289

3

913 − 24 = 889

$$\begin{array}{r} 9\;\;1\;\;3 \\ -\quad2\;\;4 \\ \hline 8\;\;8\;\;9 \end{array}$$

462 − 73 = 389

$$\begin{array}{r} 4\;\;6\;\;2 \\ -\quad7\;\;3 \\ \hline 3\;\;8\;\;9 \end{array}$$

570 − 93 = 477

$$\begin{array}{r} 5\;\;7\;\;0 \\ -\quad9\;\;3 \\ \hline 4\;\;7\;\;7 \end{array}$$

840 − 61 = 779

$$\begin{array}{r} 8\;\;4\;\;0 \\ -\quad6\;\;1 \\ \hline 7\;\;7\;\;9 \end{array}$$

355 − 76 = 279

$$\begin{array}{r} 3\;\;5\;\;5 \\ -\quad7\;\;6 \\ \hline 2\;\;7\;\;9 \end{array}$$

112 − 49 = 63

$$\begin{array}{r} 1\;\;1\;\;2 \\ -\quad4\;\;9 \\ \hline 6\;\;3 \end{array}$$

4 Write the missing digits.

(a)
$$\begin{array}{r} 3\;\;1\;\;2 \\ -\quad5\;\;5 \\ \hline 2\;\;5\;\;7 \end{array}$$

(b)
$$\begin{array}{r} 9\;\;3\;\;0 \\ -\;8\;\;7\;\;8 \\ \hline 5\;\;2 \end{array}$$

5 Subtract.

334 − 227	580 − 398	753 − 284
$\begin{array}{r} 3\;3\;4 \\ -\;2\;2\;7 \\ \hline 1\;0\;7 \end{array}$	$\begin{array}{r} 5\;8\;0 \\ -\;3\;9\;8 \\ \hline 1\;8\;2 \end{array}$	$\begin{array}{r} 7\;5\;3 \\ -\;2\;8\;4 \\ \hline 4\;6\;9 \end{array}$
I	**A**	**R**
820 − 655	843 − 386	352 − 296
$\begin{array}{r} 8\;2\;0 \\ -\;6\;5\;5 \\ \hline 1\;6\;5 \end{array}$	$\begin{array}{r} 8\;4\;3 \\ -\;3\;8\;6 \\ \hline 4\;5\;7 \end{array}$	$\begin{array}{r} 3\;5\;2 \\ -\;2\;9\;6 \\ \hline 5\;6 \end{array}$
T	**A**	**C**
736 − 389	526 − 178	440 − 397
$\begin{array}{r} 7\;3\;6 \\ -\;3\;8\;9 \\ \hline 3\;4\;7 \end{array}$	$\begin{array}{r} 5\;2\;6 \\ -\;1\;7\;8 \\ \hline 3\;4\;8 \end{array}$	$\begin{array}{r} 4\;4\;0 \\ -\;3\;9\;7 \\ \hline 4\;3 \end{array}$
N	**T**	**C**

There are bumblebees on every continent except ...
Write the letters that match the answers above to find out.

A	N	T	A	R	C	T	I	C	A
457	347	165	182	469	43	348	107	56	457

6 There are 222 townhouses and 154 condos in a new housing development.

(a) How many more townhouses are there than condos?

$$\begin{array}{r} 2\;\;2\;\;2 \\ -\;1\;\;5\;\;4 \\ \hline 6\;\;8 \end{array}$$

There are ___68___ more townhouses than condos.

(b) So far, 198 families have moved into the townhouses.
How many townhouses are still empty?

$$\begin{array}{r} 2\;\;2\;\;2 \\ -\;1\;\;9\;\;8 \\ \hline 2\;\;4 \end{array}$$

___24___ townhouses are still empty.

(c) 96 of the condos are still empty.
How many families have moved into the condos already?

$$\begin{array}{r} 1\;\;5\;\;4 \\ -\quad9\;\;6 \\ \hline 5\;\;8 \end{array}$$

___58___ families have moved into the condos.

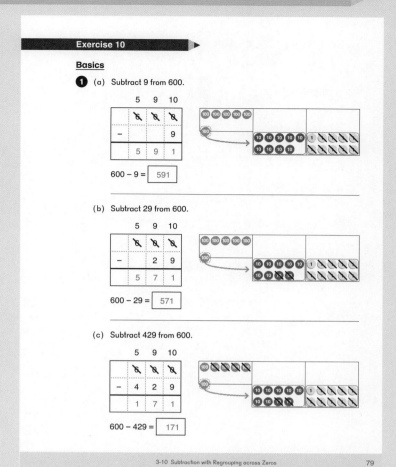

Exercise 10

Basics

1 (a) Subtract 9 from 600.

	5	9	10
	6̷	0̷	0̷
−			9
	5	9	1

600 − 9 = 591

(b) Subtract 29 from 600.

	5	9	10
	6̷	0̷	0̷
−		2	9
	5	7	1

600 − 29 = 571

(c) Subtract 429 from 600.

	5	9	10
	6̷	0̷	0̷
−	4	2	9
	1	7	1

600 − 429 = 171

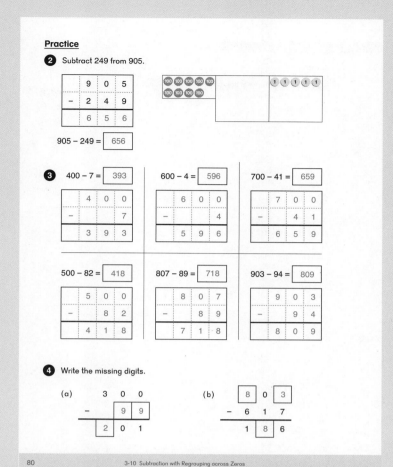

Practice

2 Subtract 249 from 905.

	9	0	5
−	2	4	9
	6	5	6

905 − 249 = 656

3 400 − 7 = 393

	4	0	0
−			7
	3	9	3

600 − 4 = 596

	6	0	0
−			4
	5	9	6

700 − 41 = 659

	7	0	0
−		4	1
	6	5	9

500 − 82 = 418

	5	0	0
−		8	2
	4	1	8

807 − 89 = 718

	8	0	7
−		8	9
	7	1	8

903 − 94 = 809

	9	0	3
−		9	4
	8	0	9

4 Write the missing digits.

(a)
	3	0	0
−		9	9
	2	0	1

(b)
	8	0	3
−	6	1	7
	1	8	6

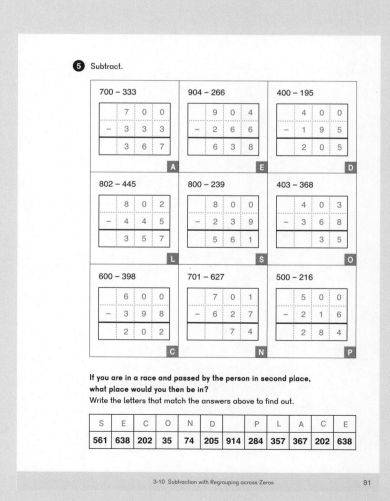

5 Subtract.

700 − 333	904 − 266	400 − 195
7 0 0 / − 3 3 3 / 3 6 7	9 0 4 / − 2 6 6 / 6 3 8	4 0 0 / − 1 9 5 / 2 0 5
A	**E**	**D**

802 − 445	800 − 239	403 − 368
8 0 2 / − 4 4 5 / 3 5 7	8 0 0 / − 2 3 9 / 5 6 1	4 0 3 / − 3 6 8 / 3 5
L	**S**	**O**

600 − 398	701 − 627	500 − 216
6 0 0 / − 3 9 8 / 2 0 2	7 0 1 / − 6 2 7 / 7 4	5 0 0 / − 2 1 6 / 2 8 4
C	**N**	**P**

If you are in a race and passed by the person in second place, what place would you then be in?
Write the letters that match the answers above to find out.

S	E	C	O	N	D		P	L	A	C	E
561	638	202	35	74	205		914	284	357	367	202	638

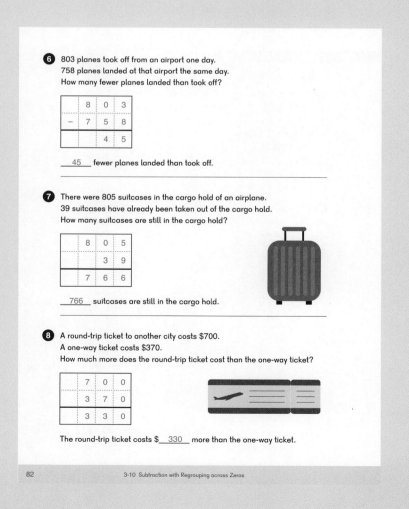

6 803 planes took off from an airport one day.
758 planes landed at that airport the same day.
How many fewer planes landed than took off?

	8	0	3
−	7	5	8
		4	5

45 fewer planes landed than took off.

7 There were 805 suitcases in the cargo hold of an airplane.
39 suitcases have already been taken out of the cargo hold.
How many suitcases are still in the cargo hold?

	8	0	5
		3	9
	7	6	6

766 suitcases are still in the cargo hold.

8 A round-trip ticket to another city costs $700.
A one-way ticket costs $370.
How much more does the round-trip ticket cost than the one-way ticket?

	7	0	0
	3	7	0
	3	3	0

The round-trip ticket costs $330 more than the one-way ticket.

Teacher's Guide 2A Chapter 3

Exercise 11

Check

1 Which two numbers below will give the greatest answer when one is subtracted from the other? Find the answer.

423	294	175	386

$$\boxed{423} - \boxed{175} = \boxed{248}$$

2 What digit is in the tens place for the value of each of the following:

Tens digit

(a) 894 – 537 $\boxed{5}$

(b) 542 – 438 $\boxed{0}$

(c) 622 – 367 $\boxed{5}$

(d) 406 – 127 $\boxed{7}$

3 Write >, <, or = in the ◯.

(a) 974 – 532 $<$ 914 – 432

(b) 704 – 263 $=$ 805 – 364

(c) 789 – 432 $>$ 325 – 61

(d) 202 + 98 $>$ 500 – 398

4

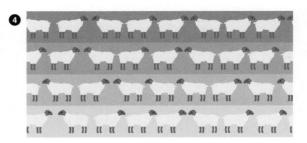

A farmer has 980 sheep.
He sheared 405 of them last week and 389 this week.

(a) How many more sheep did he shear last week than this week?

405 – 389 = 16

He sheared ___16___ more sheep last week than this week.

(b) How many sheep has he sheared so far?

405 + 389 = 794

He has sheared ___794___ sheep so far.

(c) How many sheep does he have left to shear?

980 – 794 = 186

He has ___186___ sheep left to shear.

Exercise 12

Check

1 Write one addition and one subtraction equation. Use all the numbers in the box.

467	737
375	243
710	362

$\boxed{362} + \boxed{375} = \boxed{737}$

$\boxed{710} - \boxed{467} = \boxed{243}$

2 Circle the greatest number and cross out the least number.

(a)

| 680 – 145 | 500 + 50 + 5 | 432 + 98 |

| ~~49 tens~~ | 4 hundreds 16 tens |

(b)

| 900 – 734 | 932 – 637 | 2 hundreds 60 ones |

| ~~one hundred sixty-three~~ | 189 + 76 |

(c)

| 6 hundreds – 12 ones | 4 hundreds + 17 tens | 234 + 392 |

| ~~60 + 9 + 500~~ | 956 – 382 |

---

3 Complete the number puzzles.

(a)

284	+	179	=	463
+		+		+
295	+	219	=	514
=		=		=
579	+	398	=	977

(b)

316	+	84	=	400
+		+		+
178	+	322	=	500
=		=		=
494	+	406	=	900

Challenge

4 Find the number that each shape stands for.

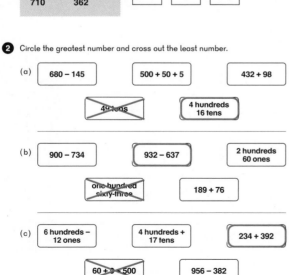

◆ + ◆ + ⬢ = 940

◆ + ⬢ = 510

 ◆ = 430

 ⬢ = 80

Hint: ◆ is the difference between 940 and 510.
Then ⬢ can be found.

Check: 430 + 430 + 80 = 940

Suggested number of class periods: 8–9

	Lesson	Page	Resources		Objectives
	Chapter Opener	p. 115	TB:	p. 93	Investigate length. Understand the need for a standard unit of measurement.
1	Centimeters	p. 116	TB: WB:	p. 94 p. 87	Understand centimeter as a unit of length. Measure the length of objects in centimeters.
2	Estimating Length in Centimeters	p. 119	TB: WB:	p. 98 p. 91	Estimate length in centimeters.
3	Meters	p. 121	TB: WB:	p. 100 p. 93	Understand meter as a unit of length. Measure the length of objects in meters.
4	Estimating Length in Meters	p. 122	TB: WB:	p. 102 p. 95	Estimate length in meters.
5	Inches	p. 124	TB: WB:	p. 104 p. 97	Understand inches as a unit of length. Measure the length of objects in inches.
6	Using Rulers	p. 126	TB: WB:	p. 106 p. 101	Understand different tools for measuring length.
7	Feet	p. 128	TB: WB:	p. 108 p. 103	Understand feet as a unit of length. Measure the length of objects in feet.
8	Practice	p. 130	TB: WB:	p. 111 p. 105	Practice measuring and estimating length. Practice adding and subtracting lengths.
	Workbook Solutions	p. 133			

In **Dimensions Math® 1B**, students learned to:

- Compare lengths directly and indirectly using nonstandard objects, such as ribbons, yarn, strips of cardboard, etc.
- Measure with non-standard units such as paper clips and cubes.

In this chapter, students will learn standard units of measurement. They will measure and compare lengths using centimeters, meters, inches, and feet.

A typical teaching sequence for measurement would follow:

In **Dimensions Math® 1**:

1. Students directly compare the length, weight, or capacity of objects.

   - The cow is taller than the cat.
   - The rock is heavier than the pebble.

2. Students indirectly compare the length, weight, or capacity of two or more objects by comparing both to a third object.

   - The height of the bookcase is the same length as the ribbon, and the chair is shorter than the ribbon. Therefore, the bookcase is taller.

3. A simple non-standard unit is used to measure objects.

   - The cat is 5 craft sticks long.
   - The rocks weigh the same as 2 linking cubes.

In **Dimensions Math® 2**:

4. A standard unit is introduced and students learn to read measurements along a scale.

   - The cat is 15 inches long.
   - The rock weighs 10 grams.

In **Dimensions Math® 3**:

5. Bigger and smaller standard units are introduced and students learn to convert different units.

At each step, students solve both calculation and word problems that build in complexity.

Much of the work in this chapter involves measuring objects in the classroom. Students will need the following tools to measure which are included as Blackline Masters:

- Centimeter rulers to 30 cm
- Inch rulers to 12 inches

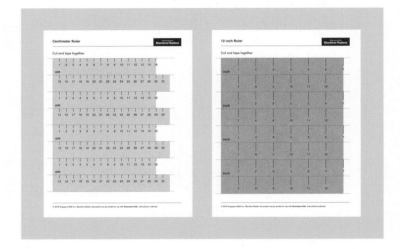

These tapes begin at 0 and do not have fractional markings on them.

Regular student rulers can be used in the first few lessons, however, students may get confused by the additional markings. Rulers with both centimeters and inches along the edges are also confusing to students and they will need to pay close attention to ensure the correct measurement is being used.

If using rulers and meter sticks that do not begin at 0, be sure to instruct students to line the objects being measured up at the 0 mark and not the end of the measuring tool. These skills will be taught formally in **Lesson 6: Using Rulers**.

All measurements are approximations. How close the approximation is to the actual measure depends on the size of the unit. In this chapter, students will only be measuring to the closest centimeter, inch, meter, or foot. They will not have to measure more precisely than about half a unit (to determine what unit the length is closest to).

It is important that students gain an intuitive feel for the units of length so they can tell when a measurement makes sense or not. A bathtub cannot be 10 m long, but a building with three floors could be 10 m high. Knowing the length of some common objects, and then estimating lengths based on that, provides students with a means to determine the reasonableness of a given measurement.

Although both the U.S. customary and the metric measurement system will be covered, students will not be converting between the two systems. In **Lesson 6**: **Using Rulers**, students will be comparing the sizes of inches and centimeters. They will learn that:

- An inch is between 2—3 centimeters long.
- A foot is about 30 centimeters long.
- A meter is much longer than a foot.
- A meter is about 3 times as long as a foot.
- An inch is about the length of a child's finger joint.
- A centimeter is about the width of a child's finger.

Materials

- 12-in rulers
- 30-cm rulers
- Centimeter cubes
- Linking cubes
- Measuring tapes
- Meter sticks
- Paper clips
- Ribbons or string, at least 3 m long
- Stickers
- Strips of paper less than 30 cm long
- Whiteboard
- Yardsticks

Blackline Masters

- 12-inch Ruler
- Centimeter Ruler

Storybooks

- *Jim and the Beanstalk* by Raymond Briggs
- *How Long or How Wide? A Measuring Guide* by Brian P. Cleary
- *How Long? (Wacky Comparisons)* by Jessica Gunderson
- *Actual Size* by Steve Jenkins
- *Measuring Penny* by Loreen Leedy
- *How Big is a Foot?* by Rolf Myller
- *Millions to Measure* by David M. Schwartz

Activities

Fewer games and activities are included in this chapter as students will be using measuring tools. The included activities can be used after students complete the **Do** questions, or anytime additional practice is needed.

Objectives

- Investigate length.
- Understand the need for a standard unit of measurement.

Lesson Materials

- *How Big is a Foot?* by Rolf Myller

Have students walk across the classroom heel-to-toe to count how many of their feet long (or wide) the classroom measures in "feet." Have an adult measure the same distance. Record all "feet" measures.

Pose the problem in the **Chapter Opener** and note that Dion and Mei have the same challenge. Discuss why the students get different numbers for the lengths.

The numerical value of what we measure depends on the unit we use. To compare, we need to use something that measures the same for everyone.

Historical measurements from the human body such as arms (cubits), hands, fingers, and legs as units of measurement could be discussed. The book *How Big is a Foot?* addresses these measurements.

Chapter 4

Length

93

Lesson 1 Centimeters

Objectives

- Understand centimeter as a unit of length.
- Measure the length of objects in centimeters.

Lesson Materials

- Paper clips of the same size
- Linking cubes
- Centimeter cubes (unit cubes from base ten blocks)
- Strips of paper less than 30 cm
- Centimeter Ruler (BLM)

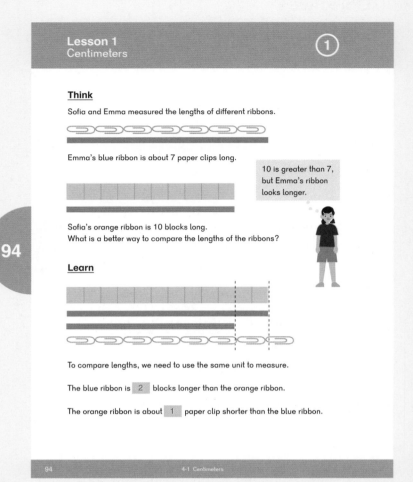

Think

Provide pairs of students with linking cubes, paper clips, and paper strips of two different lengths. Have one student measure his paper strip with the cubes and the other partner measure his paper strip with the paper clips.

Note that students' paper strips may not be exactly as long as a whole cube or paper clip. Encourage students to describe the length of their paper strips using language like:

- The paper strip is between 6 and 7 cubes long, but closer to 7.
- The paper strip is just a little more than 8 paper clips long.
- The paper strip is almost 11 cubes long.

Have students discuss why the measures of the paper strips differ. Ask them if it's possible to compare the strips with the different units (cubes and paper clips).

Learn

Have students discuss why Sofia and Emma arrived at different measurements for the ribbons.

Provide students with Centimeter Rulers (BLM) and centimeter cubes, and have them measure the length of an edge of the cube.

Note that there is no "0" on the number tapes in the textbook or on the 30-cm ruler. Students should know that the start of the tape marks the zero on the tape.

Point out that the abbreviation for centimeter is "cm."

For now, ensure that when students use the ruler, they line up the object being measured at zero.

Alex points out that the distance between tick marks is one centimeter. The number of centimeters between the end of the blue ribbon and the end of the orange ribbon is 2.

After answering the problems in the textbook, have students measure objects in the classroom with the Centimeter Rulers (BLM).

Ask students about objects they measured that were not an exact number of centimeters in length. Introduce the term "about" for estimating. For example, "The notepaper was about 8 cm long."

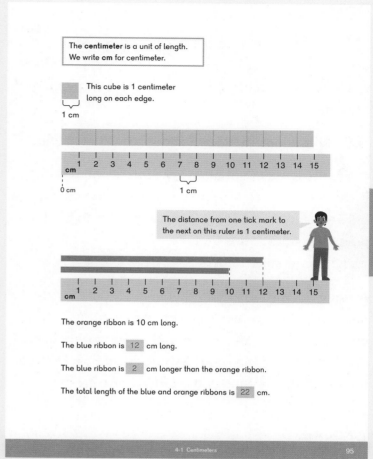

Do

1 — 2 Most objects that are measured will fall between centimeter marks, so it is important to help students see which centimeter an object's measurement is closest to.

Sofia introduces the term "about" as an estimate.

In **1**, students might also say, "It is 9 and a half cm." Point out that the red ribbon is **less** than halfway between the 9 and 10 centimeter marks. In **2**, the pipe cleaner is **more** than halfway between the 14 and 15 centimeter marks.

4 Ask students what is different in this problem compared to how they have been measuring. They should note that the crayon and eraser aren't lined up at the end of the ruler. Ask students how they can figure out the lengths of the crayon and eraser.

Activity

▲ Scavenger Hunt

Materials: Centimeter Rulers (BLM)

Provide students with a length to find in the classroom or on the playground.

Examples:

- Find something in the classroom that is about 9 cm in length.
- Find something in the classroom that is slightly more than 11 cm in length.

Exercise 1 • page 87 ▶

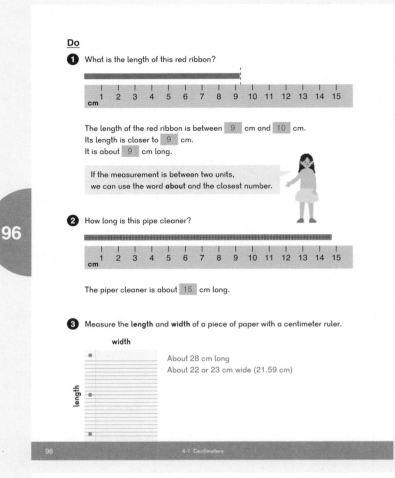

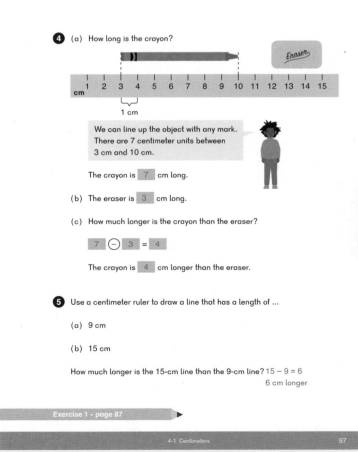

Lesson 2 Estimating Length in Centimeters

Objective

- Estimate length in centimeters.

Lesson Materials

- Centimeter Rulers (BLM) or real ruler with centimeters marked

Think

Discuss Mei's comment and then pose the first **Think** question. Provide students with Centimeter Rulers (BLM) and have them measure and record the length of their feet in centimeters.

Discuss the accuracy of the estimates.

Learn

Have students discuss the term "estimate," and Dion's and Emma's statements.

Have students measure body parts to find items that are about 1 cm and 10 cm long. Ask them to put the rulers away and estimate the length in centimeters of objects in the classroom.

Discuss the different measurements:

- My thumb is 1 cm wide. My hand from my longest finger to the base of the palm is 10 cm. My pencil is less than 1 cm wide and more than 10 cm long.
- My workbook is almost 2 hands across and more than 2 hands tall.

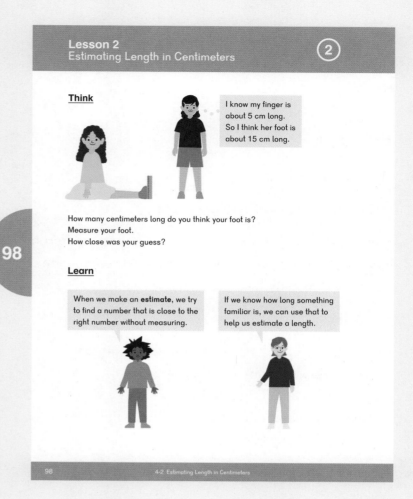

Do

Ensure students are lining the centimeter ruler up correctly.

Activities

▲ Measure Me!

Materials: Centimeter Rulers (BLM), poster board

Have students work with partners to measure different body parts. They can record their measurements on a poster and display them in the classroom.

Examples:

- My foot is about 10 cm wide and 18 cm long.
- My eyes are about 3 cm wide.
- My hair is about 3 cm tall.
- My thumb is about 2 cm wide, but my pinkie finger is about 1 cm wide.

▲ Who is Closest?

Materials: Centimeter Rulers (BLM), straightedge

This activity works best with small groups. Players take turns being the Caller or the Estimators.

For each round, the Caller chooses a length of less than 30 cm. The Estimators use a straightedge to draw a line as close as they can to the Caller's length.

Estimators then measure their lines using a 30-cm ruler. The Estimator closest to the Caller's length gets a point.

The player with the most points wins.

Exercise 2 • page 91

Do

1 Measure the distance from the tip of your thumb to the tip of your fifth finger in centimeters. Is it more or less than 10 cm?

2 Draw a line that you think is about 10 cm long. Then measure it.

3 Look for something that you think has a length of about each of the following. Then use a centimeter ruler to check.

(a) 1 cm   (b) 10 cm

(c) 15 cm   (d) 30 cm

4 Estimate the length of the following objects in centimeters. Then use a centimeter ruler to find their lengths.

(a) The length of your pencil.
Answers will vary.
(b) The length of your math textbook.
About 28 cm
(c) The width of your math textbook.
About 22 or 23 cm (21.59 cm)

99

Exercise 2 • page 91

4-2 Estimating Length in Centimeters   99

Lesson 3 Meters

Objectives

- Understand meter as a unit of length.
- Measure the length of objects in meters.

Lesson Materials

- Meter sticks
- Ribbons or strings at least 3 m long for each student

Think

Discuss the need to measure lengths that are much longer than 1 or 10 centimeters. Tell students that another unit of measurement is a meter. Provide small groups of students with meter sticks and have them find things that are about 1 meter long in the classroom or around the school.

Learn

Discuss under what circumstances length would be measured in meters rather than centimeters. Ask students:

- Why is a longer unit of measurement useful?
- Would it be handy to measure the distance from one town to another in centimeters? That would be a rather large number.

Have student look at the markings on the meter stick. Ask, "How many centimeters in one meter?"

Review textbook page 100. Point out the abbreviation of "m" for meters.

Do

❶ The length and width of the living room are not labeled. By convention, length is measured horizontally and width vertically.

Exercise 3 • page 93

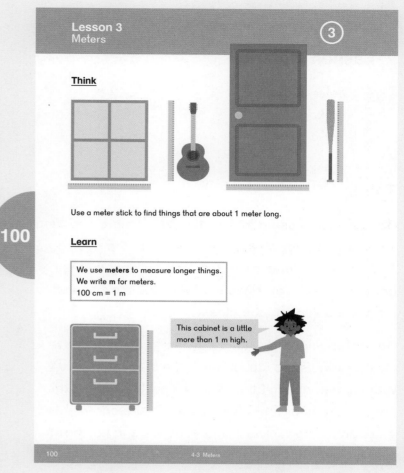

Think

Use a meter stick to find things that are about 1 meter long.

Learn

We use **meters** to measure longer things.
We write m for meters.
100 cm = 1 m

This cabinet is a little more than 1 m high.

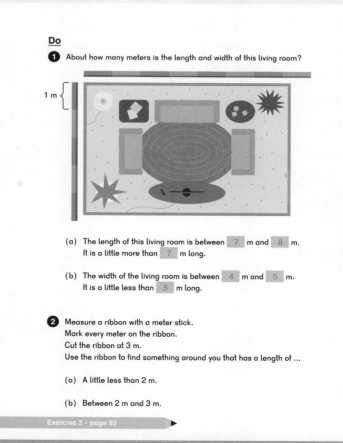

Do

❶ About how many meters is the length and width of this living room?

1 m

(a) The length of this living room is between 7 m and 8 m.
It is a little more than 7 m long.

(b) The width of the living room is between 4 m and 5 m.
It is a little less than 5 m long.

❷ Measure a ribbon with a meter stick.
Mark every meter on the ribbon.
Cut the ribbon at 3 m.
Use the ribbon to find something around you that has a length of ...

(a) A little less than 2 m.

(b) Between 2 m and 3 m.

Exercise 3 • page 93

Lesson 4 Estimating Length in Meters

Objective

- Estimate length in meters.

Lesson Materials

- Meter sticks
- Stickers

Think

Have students use a meter stick to estimate and then measure 1 meter on their body up from the floor. Then have students mark the 1-meter spot on their body with a sticker. Have students use their bodies to measure objects in the classroom.

Have students work with a partner to measure how far it is across the classroom. One student lies flat with his feet against the wall. His partner marks the spot on the floor that corresponds to the one-meter mark on his body. The student moves and lies down with his feet at the spot marked by his partner. They continue across the room until there is not enough room for the first student to lie down. They then count and write down their estimates.

Ask students, "Knowing how far it is across the classroom in one direction, can we estimate how many meters long it will be in the other direction?"

Collect student's estimates, then measure the length of the classroom.

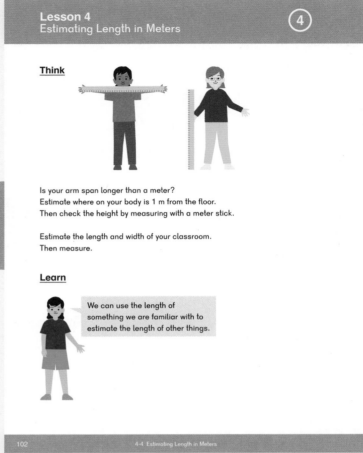

Think

Is your arm span longer than a meter?
Estimate where on your body is 1 m from the floor.
Then check the height by measuring with a meter stick.

Estimate the length and width of your classroom.
Then measure.

Learn

We can use the length of something we are familiar with to estimate the length of other things.

102

Learn

Discuss Mei's comments. Ask students to estimate objects in the classroom that are harder to measure, like the height of the doorway or a window.

Ask students whether it would make more sense to use centimeters or meters to measure:

- The height of the classroom.
- The length of a marker.
- The length of the playground.
- The height of a dog.
- The height of a giraffe.
- The length of football field.

Do

❶ Work with a partner.

Ask students, "As you get older, what will happen to the number of steps you take to measure a distance of 1 meter?"

Activity

▲ **Measure Me! — Centimeter Edition**

Materials: Large pieces of butcher paper, meter sticks, centimeter rulers

Have students work with partners to trace around and create an outline of their bodies. They can record different measurements in centimeters or meters.

"My head is 15 centimeters across. My right arm is 8 centimeters across," etc.

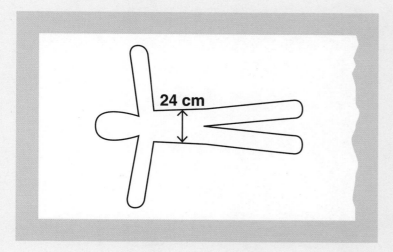

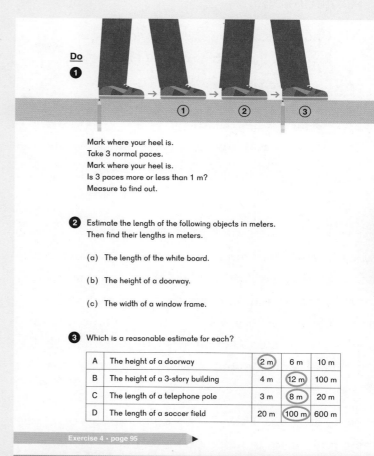

Exercise 4 • page 95

Lesson 5 Inches

Objectives

- Understand inches as a unit of length.
- Measure the length of objects in inches.

Lesson Materials

- 12-inch Ruler (BLM)
- *Millions to Measure* by David M. Schwartz

Think

Ask students what other units of measurement they know. Students will probably mention inches. Ask them to describe the length of an inch and see how familiar they are with inches. Tell students that centimeters are used in most other countries, while the United States is one of very few countries that use inches.

Provide students with 12-inch Rulers (BLM) and have them find things that can be measured in inches.

Learn

Have students note that the abbreviation for inches is "in."

Note that the period after "in" (11 in) is rarely used. If there is any confusion, simply use the whole word, "inch."

Ask if students can figure out which is longer, the centimeter or the inch.

If students are curious about the two systems of measurement, the book *Millions to Measure* is a good book to read aloud.

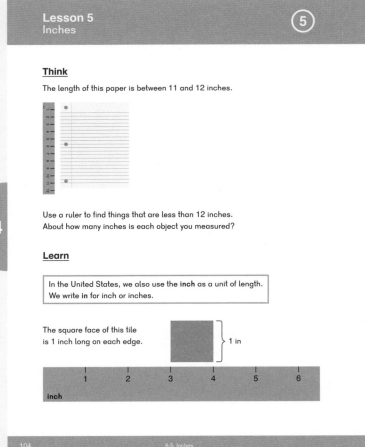

Do

1 Students should be able to compare the lengths of the pipe cleaner, crayon, and paper clip.

Activities

▲ One Inch Tall

Materials: 12-inch Ruler (BLM), the poem *One Inch Tall* by Shel Silverstein

Read the poem aloud. Have students find things in the classroom that are one inch tall using the ruler. What could they do if they were 1 in tall? What if they were 2, 3, or 5 inches tall?

▲ Who is Closest? — Inch Edition

Materials: 12-inch Rulers (BLM), straightedge

This activity works best with small groups. Players take turns being the Caller or the Estimators.

For each round, the Caller chooses a length of less than 12 inches. The Estimators use a straightedge to draw a line as close as they can to the Caller's length.

Estimators then measure their lines. The Estimator closest to the Caller's length gets a point.

The player with the most points wins.

Exercise 5 • page 97

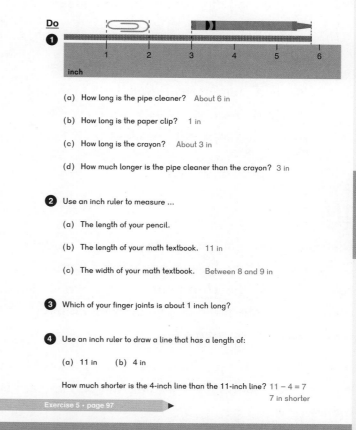

Lesson 6 Using Rulers

Objective

- Understand different tools for measuring length.

Lesson Materials

- Regular student rulers with both inch and centimeter marks
- Measuring tapes (if possible)

Think

Distribute rulers and measuring tapes to students and give them time to investigate the different rulers. Discuss the **Think** questions. Have students measure objects in the classroom with both a centimeter and an inch ruler.

Ask students, "Do you think the number we get from our measurement will be greater with centimeters or inches? Why?"

Learn

Discuss the different measurements with students. Ask them why the numerical measurements of the same object were different in inches and centimeters. Remind students of the paper clips and blocks from the first lesson in this chapter. When we use different units to measure, we get different measurements.

Lesson 6
Using Rulers   6

Think

Look at a ruler that has both centimeters and inches.

Which tick marks are for inches?
Which tick marks are for centimeters?
Where is 0 on the ruler?
Measure some things with the ruler in both centimeters and inches.

Learn

Compare 1 inch with 1 centimeter.
Which is longer?   1 inch is longer than 1 cm.

106

106

4-6 Using Rulers

Do

1 Have students look closely at the four images of rulers. Note that:

- Ruler A does not begin at zero.
- Ruler B is correct.
- Ruler C is angled, so students would not get an accurate measurement.
- Ruler D measures centimeters, not inches.

2 The goal is for students to get a feel for the centimeter being less than an inch, therefore, objects measured will have more centimeters than inches (or fewer inches than centimeters).

Activity

● **Ant Paths**

Materials: Centimeter or inch ruler, game marker (to represent the ant), die, paper

Using a blank sheet of paper as a game board, each player begins with an ant at the bottom corner. On each turn, players roll the die, then make an ant path the same length of the roll. They move their game markers (ants) and label the path.

They continue to build their paths. The first ant to cross the opposite corner wins.

▲ Create an ant maze. The longest maze that doesn't cross paths wins.

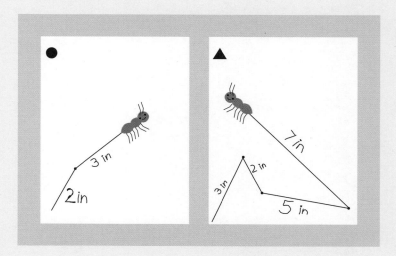

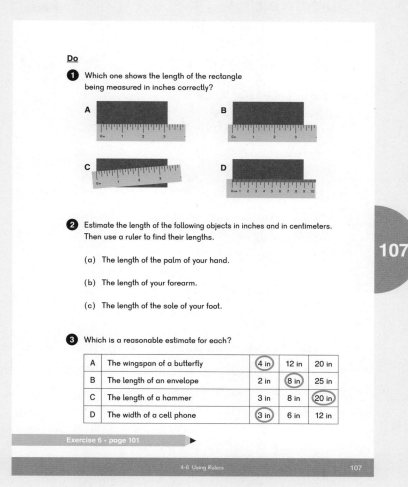

Objectives

- Understand feet as a unit of length.
- Measure the length of objects in feet.

Lesson Materials

- 12-inch rulers (regular student ruler or BLM)
- Meter sticks
- Yardsticks
- Ribbons or strings at least 3 m long for each student

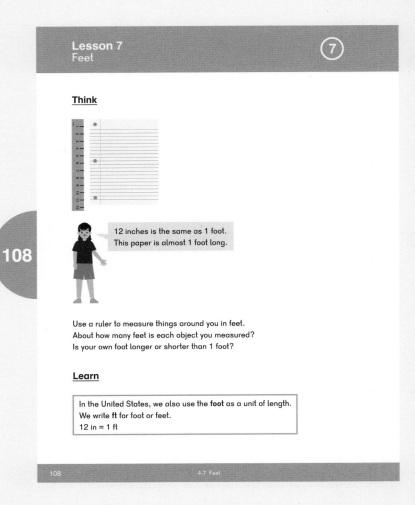

Think

Show students the meter stick and have them recall the smaller units on the meter stick (centimeters). Ask if they remember how many centimeters made one meter.

Tell students that just as there is a longer unit made up of centimeters, there is also a longer measurement made up of inches. Show students the 12-inch ruler (regular student ruler or BLM).

Have students measure things in the classroom with the 12-inch ruler.

Learn

Ask students whether it would make more sense to use inches or feet to measure:

- The height of the classroom
- The length of a marker
- The length of the playground
- The height of a dog
- The height of a giraffe
- The length of football field

Review textbook page 108. Point out the abbreviation for feet. Reiterate the fact that there are two measuring systems, metric and U.S. customary.

Do

1 Note that students are finding the number of feet for the tabletop to the nearest foot.

2 Have students measure their own desk tops.

1—**6** Have students work in pairs.

7 Dion introduces the term "yard." This is an appropriate time to compare the lengths of meter sticks and yardsticks.

Students can place three 12-inch rulers along the side of a yardstick to see that it is equal to 3 feet.

Have students estimate and measure objects in the classroom in feet and yards.

Activity

▲ Measure Me! — Foot Edition

Materials: Large pieces of butcher paper, 12-in rulers

Students can modify the drawings they completed with meters and centimeters in Lesson 4 to add measurements in feet and inches. (Perhaps use a different color.)

Alternatively, students can create a second piece showing their measurements in inches and feet.

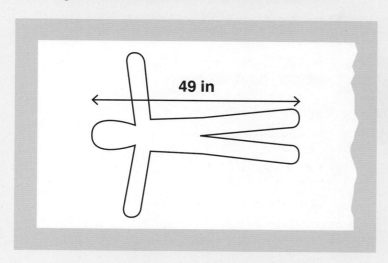

Exercise 7 • page 103

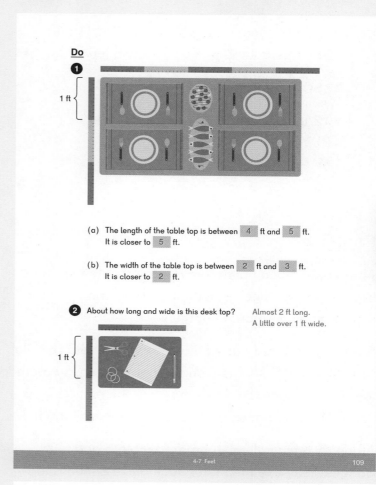

Do

1

(a) The length of the table top is between [4] ft and [5] ft. It is closer to [5] ft.

(b) The width of the table top is between [2] ft and [3] ft. It is closer to [2] ft.

2 About how long and wide is this desk top? Almost 2 ft long. A little over 1 ft wide.

3 Find a part of your body that is about 1 ft long.

4 Use a straight edge to draw a line you think is 1 ft long.

(a) Use a ruler to see how close your line is to 1 ft.
(b) About how long is it in inches?

5 Estimate the length of the following objects in feet. Then use a ruler to find their lengths in feet.

(a) Your height.
(b) The length and width of your table.
(c) The length and width of the white board.

6 Mark where your heel is.
Take 5 normal paces.
Mark where your heel is each time.
Measure the distance from one heel mark to the next in feet.
About how many feet is 1 pace?

7 Use string.

We call a length that is 3 ft long **1 yard.**

(a) Estimate a 3-ft length of string.
(b) Check how close your estimate is with a ruler.
(c) Cut a 3-ft length of string.
(d) About how long is it in inches? About 36 in
(e) About how long is it in meters? About 1 m
(f) About how long is it in centimeters? About 92 cm

Lesson 8 Practice

Objectives

- Practice measuring and estimating length.
- Practice adding and subtracting lengths.

Lesson Materials

- 12-inch rulers (regular student ruler or BLM)
- Meter sticks

Practice

After students complete the **Practice** in the textbook, have them continue measuring and estimating measurements throughout the year.

4 Students have not added 4 two-digit numbers prior to this exercise. Have them share their strategy or method for solving. Some students may line the numbers up vertically, others may add pairs of numbers together.

Students have not yet compared meters and feet.

9 Suggest they use the meter stick and 12-inch ruler (regular student ruler or BLM) to help them solve the problem.

Activity

▲ How Far?

Materials: Meter stick or measuring tape, chalk or painter's tape, index cards

Have students mark off feet or meters to make a measuring line. If outdoors, use chalk, if indoors, use painter's tape in a hallway or classroom.

Label each foot or meter on the measuring line.

Teach students how to fold a basic paper airplane. Have them start on the zero line and throw their airplanes.

Students can measure and record their throws. This measuring line can be used again in **Chapter 5: Lesson 3**.

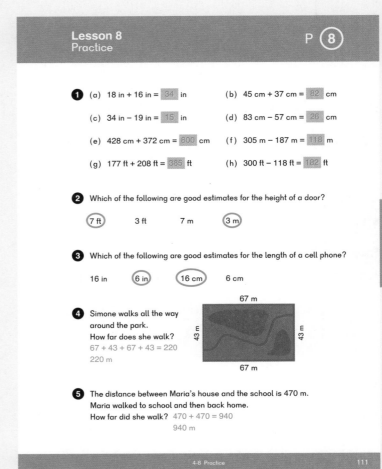

Brain Works

★ Nim

Materials: Counters

Classic Nim: Nim is a logic game that is as easy to play as tic-tac-toe, but requires more complex thinking. Once taught, students can play anytime.

Make a pile of seven counters. Two players take turns removing either one or two counters from the pile.

The player who takes the last counter is the winner.

Poison: Make a row of 3 counters, a row of 4 counters, and a row of 5 counters. Two players take turns removing 1, 2, or 3 counters from any one row.

The player forced to take the last counter eats the poison (loses).

Try changing the number of counters in each row. Does it affect the outcome? Questions to ask:

- Is there a way to always win?
- Does it matter who goes first?
- What happens if the number of counters changes?

★ How Many Cuts?

Mei has a 7-foot long pool noodle that she wants to cut into 1-foot long pieces. How many cuts does she need to make?

Emma has an old hula hoop that has a circumference (the length of the hoop) of 6 feet. She wants to cut it into 1-foot pieces for a project. How many cuts does she need to make?

Exercise 8 • page 105

Notes

Teacher's Guide 2A Chapter 4

© 2017 Singapore Math Inc.

Chapter 4 Length

Exercise 1

Basics

1 This is a centimeter ruler.
All four arrows are ___1___ cm long.

2

(a) The crayon is ___7___ cm long.

(b) The key is ___4___ cm long.

(c) The crayon is ___3___ cm longer than the key.

3

(a) The pencil is between ___9___ cm and ___10___ cm long.

(b) The craft stick is about ___14___ cm long.

Practice

4 Cut out the centimeter ruler from the back of this workbook.
Use it to measure the length of each of the objects on this page.

(a) paper clip: about ___3___ cm

(b) eraser: about ___4___ cm

(c) pipe cleaner: about ___15___ cm

(d) pen: about ___15___ cm

(e) comb: between ___11___ cm

and ___12___ cm

(f) pencil: about ___8___ cm

5 Write the length of each shaded rectangle.

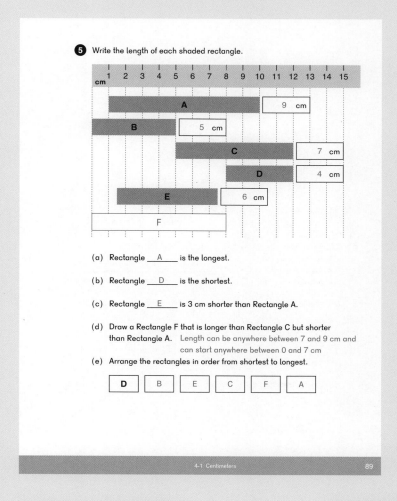

A — 9 cm
B — 5 cm
C — 7 cm
D — 4 cm
E — 6 cm
F

(a) Rectangle ___A___ is the longest.

(b) Rectangle ___D___ is the shortest.

(c) Rectangle ___E___ is 3 cm shorter than Rectangle A.

(d) Draw a Rectangle F that is longer than Rectangle C but shorter than Rectangle A. *Length can be anywhere between 7 and 9 cm and can start anywhere between 0 and 7 cm*

(e) Arrange the rectangles in order from shortest to longest.

| D | B | E | C | F | A |

6

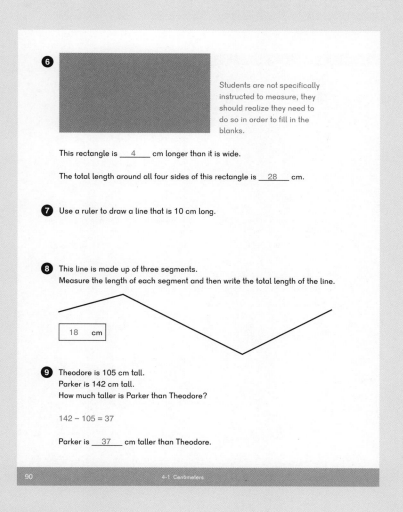

Students are not specifically instructed to measure, they should realize they need to do so in order to fill in the blanks.

This rectangle is ___4___ cm longer than it is wide.

The total length around all four sides of this rectangle is ___28___ cm.

7 Use a ruler to draw a line that is 10 cm long.

8 This line is made up of three segments.
Measure the length of each segment and then write the total length of the line.

18 cm

9 Theodore is 105 cm tall.
Parker is 142 cm tall.
How much taller is Parker than Theodore?

$142 - 105 = 37$

Parker is ___37___ cm taller than Theodore.

Teacher's Guide 2A Chapter 4

Exercise 2

Basics

1 Cut out and tape together all three centimeter rulers from the back of this workbook to make a 60-cm ruler.
Estimate the length of your thumb, hand, foot, and forearm, then measure to find the length in centimeters.   Answers will vary.

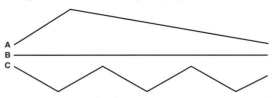

Body part		Estimated	Measured
Thumb		About _____ cm	About _____ cm
Hand		About _____ cm	About _____ cm
Foot		About _____ cm	About _____ cm
Forearm		About _____ cm	About _____ cm

2 Draw lines that you think are each of these lengths.
Then measure and write their lengths.

(a) **10 cm**   About _____ cm

(b) **6 cm**   About _____ cm

Practice

3 Look around for some objects that are less than 60 cm long.   Answers will vary.
Estimate their length first, then measure to find the length in centimeters.

Object	Estimated	Measured
	About _____ cm	About _____ cm
	About _____ cm	About _____ cm
	About _____ cm	About _____ cm
	About _____ cm	About _____ cm

4 These lines are made up of line segments.
The length of each line is the total length of the segments.

A
B
C

(a) Without measuring, arrange the lines in order from longest to shortest.

C	A	B

(b) **Line A** is 16 cm long.
Estimate the length of the other lines, then measure.

Line B is about ___15___ cm.   **Line C** is about ___17___ cm.

Exercise 3

Basics

1 Use a ribbon or cardboard strip that is 1 meter long.   Answers will vary.
Measure the lengths of the objects listed below.

Object	Length
My height	Between _____ m and _____ m
Width of a door	Between _____ m and _____ m
Height of a door	Between _____ m and _____ m
Length of a table	Between _____ m and _____ m
Width of a table	Between _____ m and _____ m

Practice

2 Measure some other lengths.   Answers will vary.

Object	Length
	Between _____ m and _____ m
	Between _____ m and _____ m
	Between _____ m and _____ m
	Between _____ m and _____ m

3 A wire is cut into two parts. The first part is 65 m long.
The second part is 32 m longer than the first part.

(a) How long is the second part?

65 + 32 = 97

The second part is ___97___ m long.

(b) How long was the wire before it was cut?

65 + 97 = 162

The wire was ___162___ m long at first.

4 The distance from the library to the school is 345 m.
The distance from the school to the playground is 429 m.

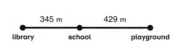

(a) How much farther is the playground than the library from the school?

429 − 345 = 84

The playground is ___84___ m farther than the library from the school.

(b) What is the total distance from the library to the school to the playground?

345 + 429 = 774

The total distance is ___774___ m.

Exercise 4

Basics

1 1 meter is the same as __100__ centimeters.

2 Use a ribbon or cardboard strip that is 1 meter long.
Look around for things that you think are about 1 m long.
Measure them to see whether they are shorter than, close to, or longer than 1 m.
Write what you measured below.
Put a check (✓) in the correct box.

Answers will vary.

Object	Shorter than 1 m	1 m	Longer than 1 m

3 Cut a string that you think is 1 m long.
Check the length of the string with a meter stick.

Practice

Answers will vary.

4 Look around for some other objects that are less than 9 m long.
Estimate their length first, then measure to find the length in meters.

Object	Estimated	Measured
	About _____ m	About _____ m
	About _____ m	About _____ m
	About _____ m	About _____ m
	About _____ m	About _____ m

5 Fill in the blanks with m or cm.

(a) The length of a bathtub is about 2 __m__.

(b) The thickness of a book is about 2 __cm__.

(c) The height of a giraffe is about 5 __m__.

(d) The depth of the water at the deep end of a pool is about 3 __m__.

(e) The wingspan of a monarch butterfly is about 10 __cm__.

(f) The height of a man is about 180 __cm__.

(g) The height of the Space Needle observation tower is about 184 __m__.

Exercise 5

Basics

1 This is an inch ruler.

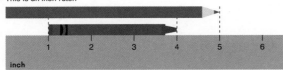

(a) The pencil is ___5___ in long.

(b) The crayon is ___3___ in long.

(c) The crayon is ___2___ in shorter than the pencil.

2

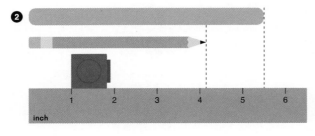

(a) The craft stick is between ___5___ in and ___6___ in long.

(b) The pencil is about ___4___ in long.

(c) The linking cube is almost ___1___ in long.

Practice

3 Cut out the inch ruler from the back of this workbook.
Use it to measure the length of each of the objects on this page.

(a) paper clip: about ___1___ in

(b) eraser: between ___1___ in
    and ___2___ in

(c) pipe cleaner: about ___6___ in

(d) pen: about ___6___ in

(e) comb: between ___4___ in
    and ___5___ in

(f) pencil: about ___3___ in

4 Cut out and tape together all three inch-rulers from the back of this workbook to make a 24-in ruler.
Estimate the length of your thumb, hand, foot, and forearm, then measure to find the length in inches.   *Answers will vary.*

Body part	Estimated	Measured
Thumb	About _____ in	About _____ in
Hand	About _____ in	About _____ in
Foot	About _____ in	About _____ in
Forearm	About _____ in	About _____ in

5 Draw lines that you think are each of these lengths.
Then measure and write their lengths.

(a) **6 in** | About _____ in

(b) **2 in** | About _____ in

6 Look around for some objects that are less than 24 in long.   *Answers will vary.*
Estimate their length first, then measure to find the length in inches.

Object	Estimated	Measured
	About _____ in	About _____ in
	About _____ in	About _____ in
	About _____ in	About _____ in
	About _____ in	About _____ in

7 Use a ruler to draw a rectangle that is 5 in long and 3 in wide on the grid.

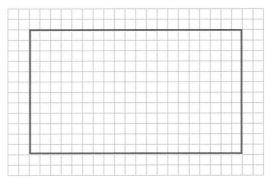

The total distance around the rectangle is ___16___ in.

Teacher's Guide 2A Chapter 4
© 2017 Singapore Math Inc.

Exercise 6

Basics

1 Check the box if the line is 3 cm long.

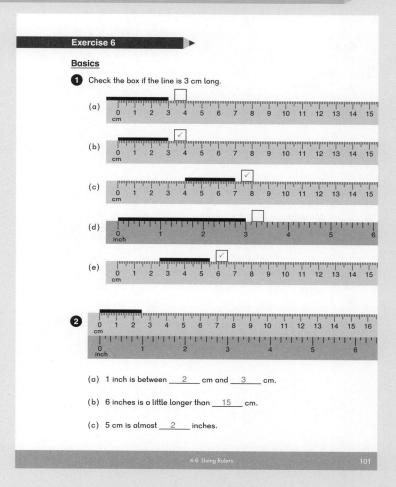

(a)

(b) ✓

(c) ✓

(d)

(e) ✓

2

(a) 1 inch is between __2__ cm and __3__ cm.

(b) 6 inches is a little longer than __15__ cm.

(c) 5 cm is almost __2__ inches.

Practice

3 Draw lines with the following lengths next to the ruler:

(a) 6 cm

(b) 4 in

4 Draw lines with the following lengths next to the ruler:

(a) 6 in

(b) 4 cm

5 Fill in the blanks with in or cm.

(a) The length of a pen is about 12 __cm__.

(b) The depth of a bookshelf is about 12 __in__.

(c) The height of a 7-year old boy is about 42 __in__.

(d) The height of a cat is about 25 __cm__.

Exercise 7

Basics

1 Use a ruler or a yard stick which shows 3 feet. Measure the lengths of the things listed below.

Answers will vary.

Object	Length
My height	Between _____ ft and _____ ft
Width of a door	Between _____ ft and _____ ft
Height of a door	Between _____ ft and _____ ft
Length of a table	Between _____ ft and _____ ft
Width of a table	Between _____ ft and _____ ft

Practice

Answers will vary.

2 Look around for some other objects that are less than 6 ft long. Estimate their length first, then measure to find the length in feet.

Object	Estimated	Measured
	About _____ ft	About _____ ft
	About _____ ft	About _____ ft
	About _____ ft	About _____ ft
	About _____ ft	About _____ ft

3 (a) 1 ft is the same as __12__ in.

(b) 1 m is about __3__ ft.

4 Fill in the blanks with ft or in.

(a) The length of a piece of paper is 11 __in__.

(b) The height of a ceiling is about 10 __ft__.

(c) The height of a stairstep is about 8 __in__.

(d) The depth of the water at the deep end of a pool is about 9 __ft__.

(e) The height of a man is about 70 __in__.

(f) The width of a soccer field is about 220 __ft__.

(g) The length of a bathtub is about 60 __in__.

(h) The wingspan of an eagle is about 6 __ft__.

5 The Prudential Tower in Boston is 907 ft tall.
The MetLife Building in New York City is 808 ft tall.
How much taller is the Prudential Tower than the MetLife Building?

907 – 808 = 99

The Prudential Tower is __99__ ft taller than the MetLife Building.

Exercise 8

Check

1 (a) Estimate, then measure the line below in centimeters.

Estimate: _____ cm

Actual length: ___8___ cm

(b) Draw a line that is 3 cm longer.

(c) Draw a line that is 4 cm shorter.

2 (a) Estimate, then measure the line below in inches.

Estimate: _____ in

Actual length: ___4___ in

(b) Draw a line that is 1 in longer.

(c) Draw a line that is 3 in shorter.

3 Write > or < in the ◯.

(a) 1 m ⊙ 1 ft [>]

(b) 1 cm ⊙ 1 in [<]

(c) 40 in ⊙ 40 cm [>]

4 (a) 489 ft + 242 ft = [731] ft

(b) 245 cm + 654 cm = [899] cm

(c) 182 in − 37 in = [145] in

(d) 231 m − 134 m = [97] m

(e) 312 cm − 68 cm = [244] cm

5 Mr. Bhakta wants to build a fence around his rectangular garden.
The garden is 124 feet long and 98 ft wide.
He will leave an opening of 12 ft for a gate on one of the longer sides
of the garden.

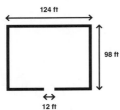

124 ft

98 ft

12 ft

(a) How much longer is the garden than it is wide?

124 − 98 = 26

The garden is ___26___ ft longer than it is wide.

(b) What is the total length of fence he needs for the side with a gate?

124 − 12 = 112

He needs ___112___ ft of fence for the side with a gate.

(c) What is the total length of fence for all four sides?
124 + 98 + 98 + 112 = 432
Students can add in steps or use mental math method, for example:
124 + 98 + 98 + 112 = 120 + 100 + 100 + 112 = 432
The total length of fence is ___432___ ft.

6 Circle the two lines below that have a total length of 12 cm.

7 cm

5 cm

6 cm

8 cm

Students should realize they
need to measure the lines.

7 4 identical blocks in a row next to each other measure 5 inches.
6 identical tiles in a row next to each other also measure 5 inches.
Which is longer, the block or the tile?

The ___block___ is longer.

Challenge

8 Four girls are comparing their heights.
Rea is shorter than Yara but taller than Nina.
Mia is shorter than Nina.
Which girl is the tallest?
Students can
draw a picture.

___Yara___ is the tallest.

9 Two yellow ribbons are the same length,
and two red ribbons are the same length.
Together, the two yellow ribbons have a length of 12 inches.
Together, the two red ribbons have a length of 20 inches.
How much longer is one red ribbon than one yellow ribbon?
6 + 6 = 12 so the yellow ribbon is 6 in.
10 + 10 = 20 so the red ribbon in 10 in.
10 − 6 = 4
One red ribbon is ___4___ inches longer than one yellow ribbon.

Students should
be able to solve
mentally.

Teacher's Guide 2A Chapter 4

Suggested number of class periods: 5–6

	Lesson	Page	Resources		Objectives
	Chapter Opener	p. 143	TB:	p. 113	Measure and compare the weight of objects in grams.
1	Grams	p. 144	TB: WB:	p. 114 p. 109	Understand a gram as a unit of weight. Measure and compare the weight of objects in grams.
2	Kilograms	p. 147	TB: WB:	p. 118 p. 113	Understand a kilogram as a unit of weight. Measure and compare the weight of objects in kilograms.
3	Pounds	p. 149	TB: WB:	p. 121 p. 115	Understand a pound as a unit of weight. Measure and compare the weight of objects in pounds. Compare one kilogram to one pound.
4	Practice	p. 151	TB: WB:	p. 124 p. 117	Practice topics from the chapter.
	Review 1	p. 153	TB: WB:	p. 126 p. 119	Review content from Chapter 1 through Chapter 5.
	Workbook Solutions	p. 155			

In **Dimensions Math® KA**, students learned to:

- Understand the terms "heavy," "heavier than," "light," and "lighter than."
- Compare and measure weights directly and indirectly using non-standard units.

In this chapter, students extend their knowledge of measuring weight from using non-standard units to using the standard units of grams, kilograms, and pounds.

Much of the work involves weighing objects in the classroom. Students need the following tools:

- Balance scale, one per group of 3 to 4

- Weights: 1 g, 10 g, 100 g, 500 g, and 1 kg

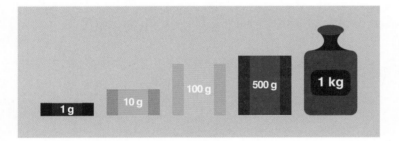

- Beans (and bags) to make 1-lb weights

Students will learn to estimate the weight of grams, kilograms, and pounds. Although both metric and U.S. customary measuring systems are covered, students will not convert between the two systems. In **Lesson 3: Pounds**, students compare the weight of one kilogram to that of one pound.

One kilogram is about 2.2 pounds. One pound is about 0.45 kilograms. Students might realize that a kilogram weighs roughly twice as much as a pound. This will be a good estimate to use in determining whether a weight in kilograms makes sense if you are used to pounds. It is reasonable for a second grade student to weigh 22 kg, which is about 44 lb.

Weight and mass are different concepts, though in normal life we don't worry about this distinction. The metric system measures mass but our standard English units measure weight.

Grams and kilograms are a measure of mass. Mass is the measure of the object's resistance to acceleration when a net force is applied, and is related to the amount of matter in that object. It does not depend on gravity. Pounds and ounces are a measure of weight. Weight is the measure of the force of gravity. The Earth pulls all objects toward its center, and this is true regardless of the motion of the object.

In this chapter, students weigh objects to find either their mass or weight, but are not required to understand the two concepts. The term "weight" is used at this level.

Students should understand that the mass or weight of an object is not related to its size (volume or capacity). A pillow may be lighter than a rock even though it may be larger than the rock.

Materials

- 1-kg weights
- 1-lb weights
- Pennies (in rolls and loose)
- Balance scales showing kg and lb
- 1-g weights
- Linking cubes
- Objects weighing about 1 kg
- Objects weighing about 1 lb
- Objects weighing about 1 g
- Classroom objects to weigh, including pens and pencils
- Zipper bags of various sizes
- Dried beans
- Whiteboards

Blackline Masters

- Shisima Game Board

Storybooks

- *Measuring Penny* by Loreen Leedy
- *Mighty Maddie* by Stuart J. Murphy
- *Weight* by Henry Pluckrose
- *Me and the Measure of Things* by Joan Sweeney

Activities

Fewer games and activities are included in this chapter as students will be using measuring tools. The included activities can be used after students complete the **Do** questions, or anytime additional practice is needed.

Notes

Chapter Opener

Objective

- Measure and compare the weight of objects in grams.

Materials

- Balance scale
- Linking cubes
- Objects to weigh such as a pencil, attribute bears, staples, etc.

In **Dimensions Math® KA**, students used a balance scale to compare weights using non-standard items, such as a toy car that weighed the same as 10 linking cubes. As weight was not covered in grade 1, this lesson familiarizes students with a balance scale, and shows how to weigh and compare objects using non-standard units (such as linking cubes).

For example:

- My pencil weighs the same as 5 linking cubes.
- The marker weighs the same as 7 linking cubes.
- The marker weighs more than the pencil. The pencil weighs less than the marker.

Show students how to tare the balance, that is, set the slider so that the two pans balance before any weights are put on them.

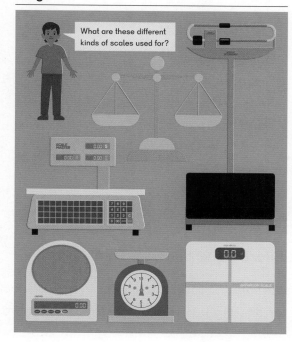

Chapter 5

Weight

What are these different kinds of scales used for?

113

Lesson 1 Grams

Objectives

- Understand a gram as a unit of weight.
- Measure and compare the weight of objects in grams.

Materials

- Balance scale
- Gram weights
- Pens and pencils to weigh
- Paper clips and other objects that weigh under one kilogram

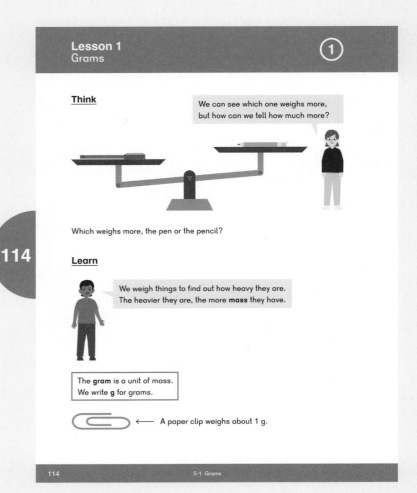

Think

Pose Emma's question from **Think**.

Ask students to recall the lesson in the previous chapter on length in which they measured strips of paper with linking cubes and paper clips. Ask if they remember why it was difficult to measure the paper strips with different objects.

Tell them that just as we measure length with units like inches and centimeters, we also have standard units for weight.

Provide groups of students balance scales and gram weights. Have them find the weight of a pencil and pen in grams. Ask how many grams heavier one object is than the other.

Learn

Have students discuss Alex's comments and the term "mass." Introduce the term "gram" and its abbreviation, and the 1-g weight.

Have students find the weight of different objects in the classroom.

After discussing the subtraction equation, have students compare the weight of a pen and a pencil by placing them on opposite sides of the balance scale. Have students add one gram weights to the pencil side until the two sides are even.

This can also be shown with a comparison bar model:

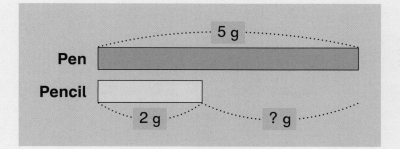

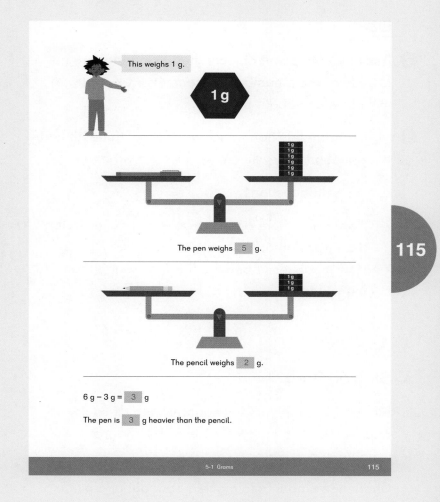

Do

3 Ask students about objects they weigh that are not exactly a whole number of grams. For example, "A nickel weighs about 5 grams."

5 — **6** Bar models can be drawn to help students find the equation they will use to solve the problem.

Extend with questions:

- How much more does the tape dispenser weigh than the stapler?
- How much less does the thermos weigh than the water?

Activity

▲ Sink or Float?

Materials: Aluminum foil squares with sides of approximately 6 in, centimeter cubes or gram weights, tub of water

Challenge students to make a boat out of a piece of aluminum foil. Have them estimate how many grams their boats can hold before sinking.
(1 cm cube = 1 g)

Have students test their boats to see which can hold the most weight.

Exercise 1 • page 109

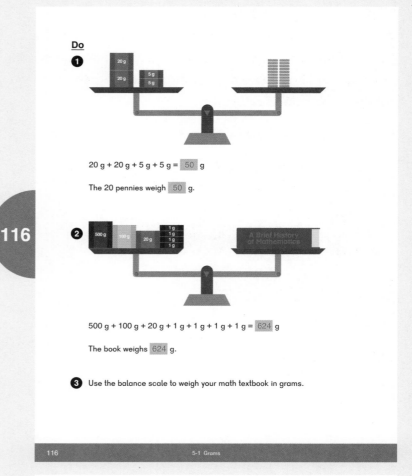

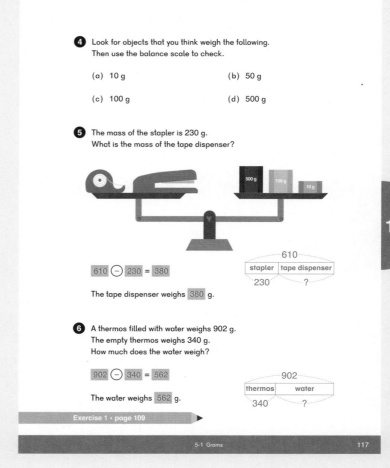

Lesson 2 Kilograms

Objectives

- Understand a kilogram as a unit of weight.
- Measure and compare the weight of objects in kilograms.

Materials

- Balance scales
- 1-kg weights
- Objects that weigh about 1 kilogram
- Zipper bags and dried beans
- 400+ real pennies (can use some rolls of 50 and some loose)
- Bathroom scale showing kilograms

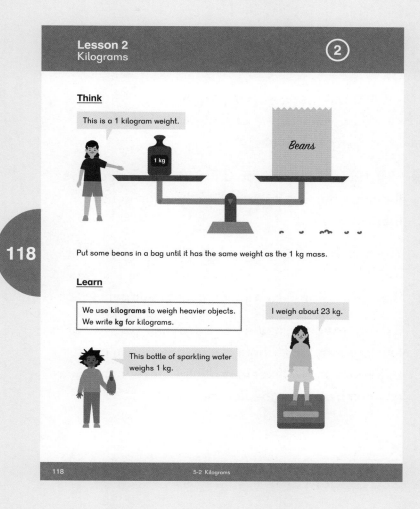

Think

Discuss the need to measure weights that are much heavier than 10 or 20 grams. Tell students that another unit of measurement is the kilogram.

Provide groups of students with a balance scale and a 1-kg weight. As there will probably not be enough 1-kg weights for all groups to use one at the same time, students can take turns making a 1-kg weight by filling a bag with beans until it balances the kilogram weight. Label the bag "1 kilogram."

Learn

Introduce the term "kilogram" and the abbreviation. Students will be weighing more objects in the next portion of the lesson.

Inquisitive students might want to know how many grams are in a kilogram (1,000).

Do

1 Note that 1 kilogram is about:

- 400 U.S. pennies (not play money)
- 40 crayons
- 4 spiral notebooks

1—**2** Have students record their answers and save them for the next lesson when they can compare them to pounds.

The activities encourage students to get a feel for weights. Does their backpack weigh more or less than 1 kilogram? How much might their pets weigh?

5 Extend by having students guess their body weights in kilograms. Have them weigh themselves on a bathroom scale. If scale is digital, have them ignore any numbers after the decimal point.

Activity

▲ Scavenger Hunt

Materials: Scales, kilogram weights

Challenge students to find something in the classroom that weighs 1 kilogram.

Exercise 2 • page 113 ▶

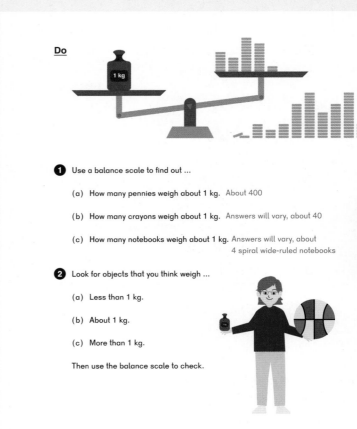

Do

1 Use a balance scale to find out ...

(a) How many pennies weigh about 1 kg. About 400

(b) How many crayons weigh about 1 kg. Answers will vary, about 40

(c) How many notebooks weigh about 1 kg. Answers will vary, about 4 spiral wide-ruled notebooks

2 Look for objects that you think weigh ...

(a) Less than 1 kg.

(b) About 1 kg.

(c) More than 1 kg.

Then use the balance scale to check.

119

5-2 Kilograms   119

3 Find something that you think weighs the following.
Then use a weighing scale to check.

(a) 3 kg   (b) 5 kg

4 Use the weights to find how many grams is the same as 1 kilogram.

1 kg = 1,000 g

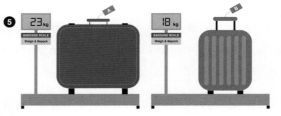

5 23 kg   18 kg

(a) How much more does suitcase A weigh than suitcase B?

23 ⊖ 18 = 5

Suitcase A weighs 5 kg more than suitcase B.

(b) How much do they weigh altogether?

23 ⊕ 18 = 41

They weigh 41 kg altogether.

Exercise 2 - page 113 ▶

120   5-2 Kilograms

Lesson 3 Pounds

Objectives

- Understand a pound as a unit of weight.
- Measure and compare the weight of objects in pounds.
- Compare one kilogram to one pound.

Lesson Materials

- Balance scales
- 1-lb and 1-kg weights
- Objects that weigh about one pound
- Zipper bags and dried beans
- 200+ real pennies (can use some rolls of 50 and some loose)

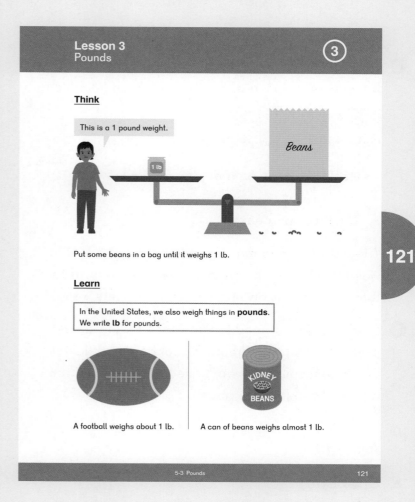

Think

Tell students that just as there are different systems for measuring length, there are also different systems for measuring weight. Grams and kilograms are metric measurements of weight. Ask students if they know other terms for measuring weight. Students may know pounds.

Provide groups of students with balance scales and 1-pound weights. If there are not enough weights, have students make them by filling bags with 1 pound of beans each and labeling them "1 pound."

Learn

Introduce the term "pound" and the abbreviation. Students will weigh more objects in the next portion of the lesson.

Students may be curious about why the word "pound" is abbreviated "lb." The abbreviation comes from the Roman unit of measure, *libra pondo*, meaning "a pound by weight."

Do

❶ — ❷ Students can use their recording sheet from the previous lesson and note how much each item weighs in pounds.

Note that 1 pound is about:

- 182 U.S. pennies (not play money)
- 18 crayons
- 2 spiral notebooks

❷ The activities encourage students to get a feel for weights. Do their lunchboxes weigh more or less than 1 pound? How much might their pets weigh?

❸ Students do not need to know that 1 kilogram is about 2.2 pounds. "About 2 pounds" is sufficient. The objective is for them to see that the kilogram weighs more that the pound.

Activity

▲ Shot Put

Materials: 1-kg bags of beans, 1-lb bags of beans, extra zipper bags in quart and gallon sizes, chalk or painter's tape, index cards

Create a measuring line in feet outside or in the hallway like the one made for the paper airplane activity **How Far?** in **Chapter 4: Lesson 8**.

Double (or triple) bag the 1-kg and 1-lb bags of beans.

Have students measure how many feet they can throw the bean bags. Does the weight of the bag affect how far they can throw it?

Create a 5-pound bean bag. Can students throw 5 pounds as far as they can throw 1 pound?

Exercise 3 • page 115

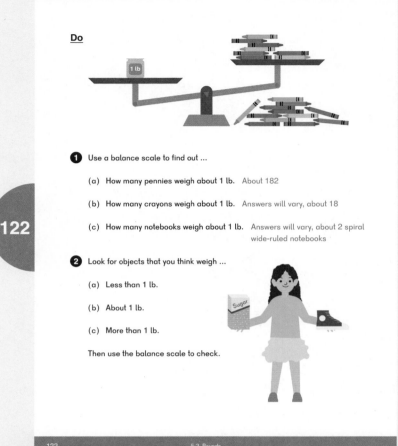

Do

❶ Use a balance scale to find out ...

(a) How many pennies weigh about 1 lb.   About 182

(b) How many crayons weigh about 1 lb.   Answers will vary, about 18

(c) How many notebooks weigh about 1 lb.   Answers will vary, about 2 spiral wide-ruled notebooks

❷ Look for objects that you think weigh ...

(a) Less than 1 lb.

(b) About 1 lb.

(c) More than 1 lb.

Then use the balance scale to check.

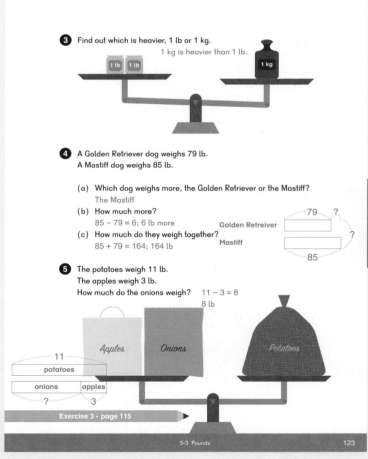

❸ Find out which is heavier, 1 lb or 1 kg.
1 kg is heavier than 1 lb.

❹ A Golden Retriever dog weighs 79 lb.
A Mastiff dog weighs 85 lb.

(a) Which dog weighs more, the Golden Retriever or the Mastiff?
The Mastiff
(b) How much more?
85 − 79 = 6; 6 lb more   Golden Retriever   79   ?
(c) How much do they weigh together?   Mastiff   ?
85 + 79 = 164; 164 lb   85

❺ The potatoes weigh 11 lb.
The apples weigh 3 lb.
How much do the onions weigh?   11 − 3 = 8
8 lb

11
potatoes
onions   apples
?   3

Exercise 3 • page 115

Lesson 4 Practice

Objective

- Practice topics from the chapter.

Practice

After students complete the **Practice** in the textbook, have them continue measuring and estimating length and weight throughout the year.

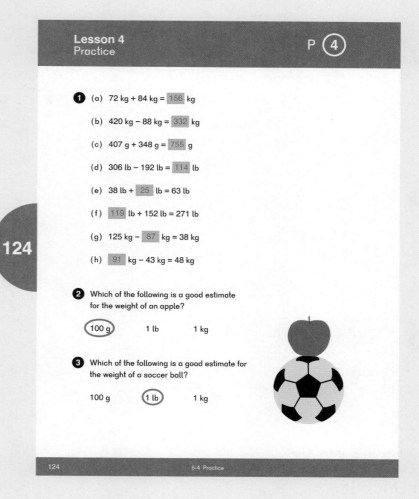

1 (a) 72 kg + 84 kg = 156 kg

(b) 420 kg − 88 kg = 332 kg

(c) 407 g + 348 g = 755 g

(d) 306 lb − 192 lb = 114 lb

(e) 38 lb + 25 lb = 63 lb

(f) 119 lb + 152 lb = 271 lb

(g) 125 kg − 87 kg = 38 kg

(h) 91 kg − 43 kg = 48 kg

124

2 Which of the following is a good estimate for the weight of an apple?

(100 g)   1 lb   1 kg

3 Which of the following is a good estimate for the weight of a soccer ball?

100 g   (1 lb)   1 kg

124   5-4 Practice

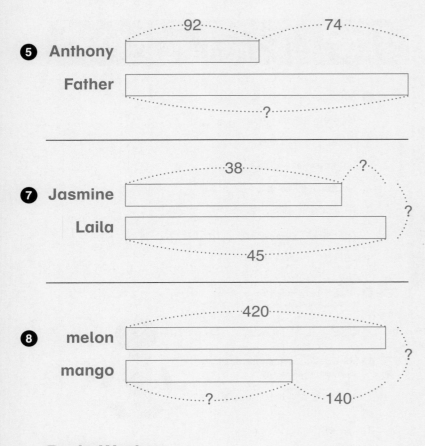

⑤ Anthony ··········92·········· ··········74··········

Father ?

⑦ Jasmine ·······38······· ?

Laila ?

·······45·······

⑧ melon ········420········

mango ?

? ···140···

Brain Works

★ Mobiles

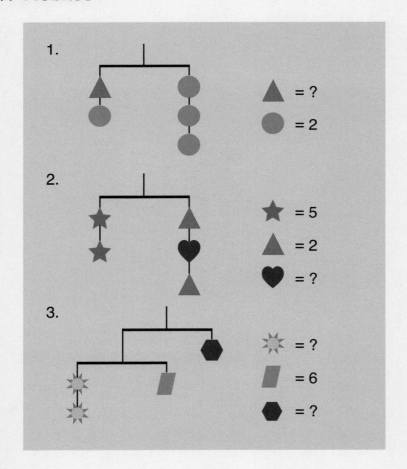

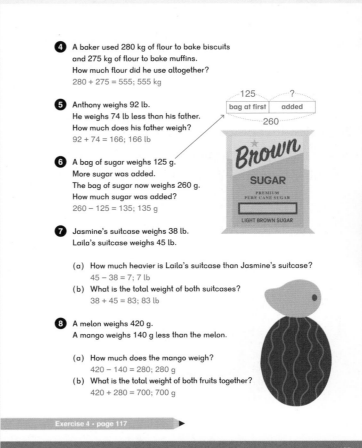

④ A baker used 280 kg of flour to bake biscuits and 275 kg of flour to bake muffins. How much flour did he use altogether?
280 + 275 = 555; 555 kg

⑤ Anthony weighs 92 lb. He weighs 74 lb less than his father. How much does his father weigh?
92 + 74 = 166; 166 lb

⑥ A bag of sugar weighs 125 g. More sugar was added. The bag of sugar now weighs 260 g. How much sugar was added?
260 − 125 = 135; 135 g

⑦ Jasmine's suitcase weighs 38 lb. Laila's suitcase weighs 45 lb.

(a) How much heavier is Laila's suitcase than Jasmine's suitcase?
45 − 38 = 7; 7 lb
(b) What is the total weight of both suitcases?
38 + 45 = 83; 83 lb

⑧ A melon weighs 420 g. A mango weighs 140 g less than the melon.

(a) How much does the mango weigh?
420 − 140 = 280; 280 g
(b) What is the total weight of both fruits together?
420 + 280 = 700; 700 g

Exercise 4 • page 117

5-4 Practice 125

Exercise 4 • page 117

Review 1

Objective

- Review topics from Chapter 1 through Chapter 5.

Use this cumulative review as needed to practice and reinforce content and skills from the first five chapters.

Activities

▲ Cryptarithmetic Puzzles

Each letter stands for a number 0–9. The first three have multiple solutions. Challenge students with, "How many solutions can you find?"

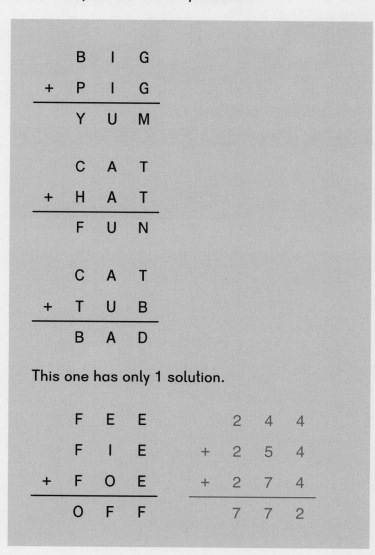

```
    B  I  G
 +  P  I  G
 ----------
    Y  U  M


    C  A  T
 +  H  A  T
 ----------
    F  U  N


    C  A  T
 +  T  U  B
 ----------
    B  A  D
```

This one has only 1 solution.

```
    F  E  E              2  4  4
    F  I  E           +  2  5  4
 +  F  O  E           +  2  7  4
 ----------           ----------
    O  F  F              7  7  2
```

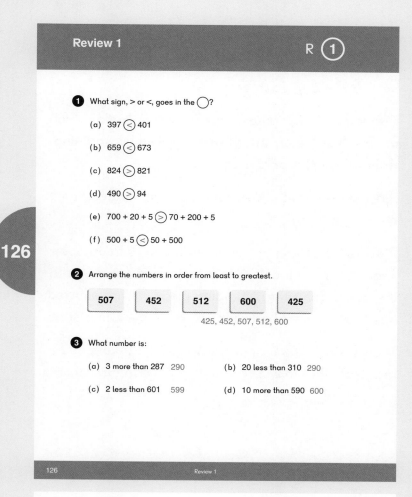

Review 1 R ①

1 What sign, > or <, goes in the ◯?

(a) 397 ⓒ 401

(b) 659 ⓒ 673

(c) 824 ⓢ 821

(d) 490 ⓢ 94

(e) 700 + 20 + 5 ⓢ 70 + 200 + 5

(f) 500 + 5 ⓒ 50 + 500

126

2 Arrange the numbers in order from least to greatest.

| 507 | 452 | 512 | 600 | 425 |

425, 452, 507, 512, 600

3 What number is:

(a) 3 more than 287 290   (b) 20 less than 310 290

(c) 2 less than 601 599   (d) 10 more than 590 600

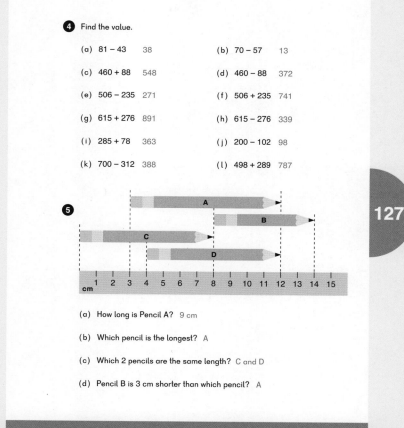

4 Find the value.

(a) 81 − 43 38   (b) 70 − 57 13

(c) 460 + 88 548   (d) 460 − 88 372

(e) 506 − 235 271   (f) 506 + 235 741

(g) 615 + 276 891   (h) 615 − 276 339

(i) 285 + 78 363   (j) 200 − 102 98

(k) 700 − 312 388   (l) 498 + 289 787

127

5

(a) How long is Pencil A? 9 cm

(b) Which pencil is the longest? A

(c) Which 2 pencils are the same length? C and D

(d) Pencil B is 3 cm shorter than which pencil? A

▲ Shisima

Materials: Shisima Game Board (BLM), 6 game markers — 3 of each color

This logic game comes from Kenya. *Shisima*, which means "source of water," was inspired by watching water bugs dart across the water.

Players begin with their 3 markers on the dots, as in the image below.

Players take turns moving their markers to an open space. A move must be to an adjacent corner, or to the center (shisima). Jumping pieces is not allowed, and there cannot be two markers on the same space.

The winner is the first player to get 3 markers in a row. Note that to win, 1 marker will be in the shisima.

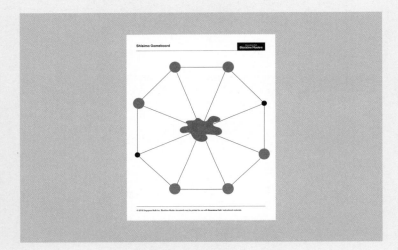

Exercise 5 • page 119

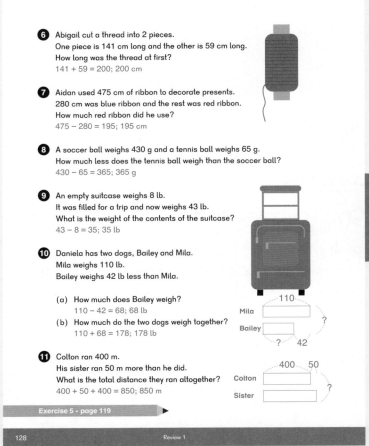

6 Abigail cut a thread into 2 pieces.
One piece is 141 cm long and the other is 59 cm long.
How long was the thread at first?
141 + 59 = 200; 200 cm

7 Aidan used 475 cm of ribbon to decorate presents.
280 cm was blue ribbon and the rest was red ribbon.
How much red ribbon did he use?
475 − 280 = 195; 195 cm

8 A soccer ball weighs 430 g and a tennis ball weighs 65 g.
How much less does the tennis ball weigh than the soccer ball?
430 − 65 = 365; 365 g

9 An empty suitcase weighs 8 lb.
It was filled for a trip and now weighs 43 lb.
What is the weight of the contents of the suitcase?
43 − 8 = 35; 35 lb

10 Daniela has two dogs, Bailey and Mila.
Mila weighs 110 lb.
Bailey weighs 42 lb less than Mila.

   (a) How much does Bailey weigh?
        110 − 42 = 68; 68 lb
   (b) How much do the two dogs weigh together?
        110 + 68 = 178; 178 lb

11 Colton ran 400 m.
His sister ran 50 m more than he did.
What is the total distance they ran altogether?
400 + 50 + 400 = 850; 850 m

Exercise 5 • page 119

128   Review 1

Chapter 5 Weight

Exercise 1

Basics

1 Use a balance scale and some gram weights.
Weigh the following objects and write the mass in grams.
Answers will vary.

Object	Mass
A pencil	_____ grams
A mug	_____ grams
5 crayons	_____ grams

2 Look for some items that you think weigh about 200 g.
Weigh them to see if they are less than or more than 200 g.
Write what you weighed below.
Put a check (✔) in the correct box.
Answers will vary.

Object	Lighter than 200 g	Heavier than 200 g

Practice

3 Look around for some objects that you think weigh less than 1,000 g.
Estimate their mass first, then weigh to find the mass in grams.
Answers will vary.

Object	Estimated	Weighed
	About _____ g	About _____ g
	About _____ g	About _____ g
	About _____ g	About _____ g
	About _____ g	About _____ g

4 How much does each item weigh?

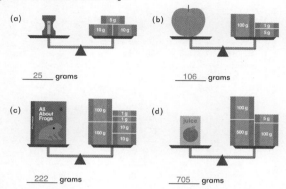

(a) ___25___ grams

(b) ___106___ grams

(c) ___222___ grams

(d) ___705___ grams

5 Fill in the blanks.

(a) The pear weighs ___127___ g.

(b) The banana weighs ___165___ g.

(c) The total weight of the fruits is ___292___ g.
127 + 165 = 292

(d) The banana weighs ___38___ g more than the pear.
165 − 127 = 38

6 A can of tuna weighs 199 g.
A can of tomato sauce weighs 425 g.

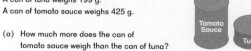

(a) How much more does the can of tomato sauce weigh than the can of tuna?

425 − 199 = 226

The can of tomato sauce weighs ___226___ g more than the can of tuna.

(b) What do both cans weigh together?

425 + 199 = 624

Both cans weigh ___624___ g together.

7 How much weight needs to be added to the left side to balance the right side?

530 g   625 g

625 − 530 = 95

___95___ g needs to be added to the left side.

8 How much does the avocado weigh?

avocado + 15 g = 230
230 − 15 = 215

The avocado weighs ___215___ g.

Challenge

A will balance if the 20 g is removed and 50 g replaced with 30 g.
B will balance if the 70 g is removed and 300 g replaced with 230 g.

9 How much 200 g is the same for both. than Present B?

230 − 30 = 200
Or: A weighs 230, B weighs 430, difference is 200.

Present A weighs ___200___ g less than Present B.

Teacher's Guide 2A Chapter 5

Exercise 2

Basics

1 Look for some items that you think weigh about 1 kilogram.
Use a 1-kg mass and a balance scale to see if they are lighter than, close to, or heavier than 1 kg.
Write what you weighed below.
Put a check (✓) in the correct box.
Answers will vary.

Object	Lighter than 1 kg	1 kg	Heavier than 1 kg

Practice

2 Use a weighing scale that measures in kilograms.
Find some objects that are heavier than 1 kg.
Estimate their mass first, then weigh to find the mass in kilograms.
Answers will vary.

Object	Estimated	Weighed
	About _____ kg	About _____ kg
	About _____ kg	About _____ kg
	About _____ kg	About _____ kg
	About _____ kg	About _____ kg

5-2 Kilograms   113

---

3 Fill in the blanks with g or kg.

(a) A newborn kitten weighs about 100 __g__.

(b) A man weighs about 80 __kg__.

(c) A cell phone weighs about 120 __g__.

(d) A wide-screen computer monitor weighs about 4 __kg__.

(e) A 1-L bottle of water weighs about 1 __kg__.

(f) A quarter weighs about 6 __g__.

(g) A bicycle weighs about 15 __kg__.

4 A bakery ordered 120 kg of flour and 95 kg of sugar.

(a) How much more flour than sugar did it order?

120 − 95 = 25

It ordered __25__ kg more flour than sugar.

(b) The bakery used 87 kg of the flour so far.
How much flour is left?

120 − 87 = 33

__33__ kg of flour is left.

114   5-2 Kilograms

Exercise 3

Basics

1 Look for some items that you think weigh about 1 pound.
Use a 1-lb weight and a balance scale to see if they are lighter than, close to, or heavier than 1 lb.
Write what you weighed below.
Put a check (✓) in the correct box.
Answers will vary.

Object	Lighter than 1 lb	1 lb	Heavier than 1 lb

Practice

2 Use a weighing scale that measures in pounds.
Find some objects that are heavier than 1 lb.
Estimate their weight, then weigh to find the mass in pounds.
Answers will vary.

Object	Estimated	Weighed
	About _____ lb	About _____ lb
	About _____ lb	About _____ lb
	About _____ lb	About _____ lb
	About _____ lb	About _____ lb

5-3 Pounds   115

---

3 Circle the most reasonable weight for each of the following.

(a) A bicycle

3 lb   (30 lb)   100 lb

(b) A snowmobile

6 lb   60 lb   (600 lb)

(c) A laundry basket filled with clothes

(8 lb)   80 lb   800 lb

Challenge

4 The weight of 2 identical washing machines and a dryer, and 1 washing machine and a dryer are shown below.
How much does the dryer weigh?

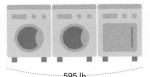
595 lb

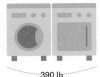
390 lb

595 − 390 = 205 (weight of washing machine)
390 − 205 = 185 (weight of dryer)

The dryer weighs __185__ lb.

116   5-3 Pounds

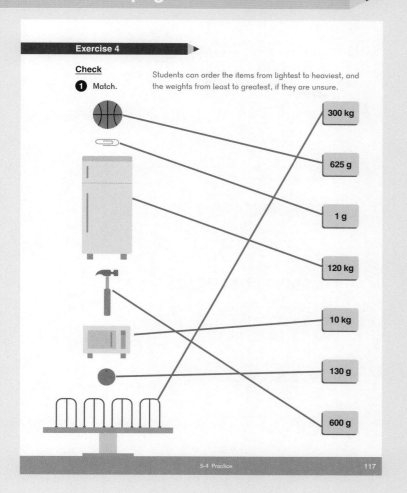

Check

1 Match.

Students can order the items from lightest to heaviest, and the weights from least to greatest, if they are unsure.

300 kg

625 g

1 g

120 kg

10 kg

130 g

600 g

5-4 Practice   117

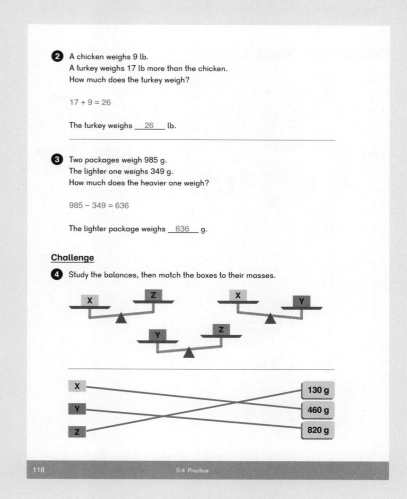

2 A chicken weighs 9 lb.
A turkey weighs 17 lb more than the chicken.
How much does the turkey weigh?

17 + 9 = 26

The turkey weighs ___26___ lb.

3 Two packages weigh 985 g.
The lighter one weighs 349 g.
How much does the heavier one weigh?

985 − 349 = 636

The lighter package weighs ___636___ g.

Challenge

4 Study the balances, then match the boxes to their masses.

X
Y
Z

130 g
460 g
820 g

118   5-4 Practice

Teacher's Guide 2A Chapter 5

Exercise 5

Check

1 Write the numbers.

(a) 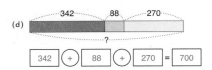 889

(b) 406

(c) 580

2 Write the numbers.

5 hundreds, 4 tens, 8 ones	548
8 ones, 6 hundreds	608
900 + 3 + 70	973
2 more than five hundred forty-nine	551
10 less than 8 hundreds and 6 ones	796
400 more than 5 tens 5 ones	455

3 Complete the number patterns.

(a) **746** **766** 786 806 826 **846**

(b) **365** **363** 361 359 357 **355**

4 Write the greatest number and the least number using all three digits.

	greatest number	least number
1, 9, 4	941	149
6, 0, 7	706	607
3, 8, 3	833	338

5 Complete the equation using the information in the model.
Find the missing numbers.

(a)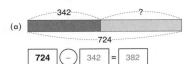

342   ?
724

724 $-$ 342 $=$ 382

(b)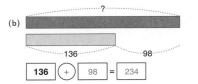

?
136   98

136 $+$ 98 $=$ 234

(c)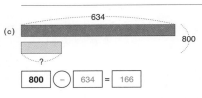

634
800
?

800 $-$ 634 $=$ 166

(d) 342   88   270
?

342 $+$ 88 $+$ 270 $=$ 700

6 Write the missing digits.

(a)
```
    3  2  6
 +  3  7  4
    7  0  0
```

(b)
```
    6  0  8
 +     9  5
    7  0  3
```

(c)
```
    7  0  0
 -        2  8
       6  7  2
```

(d)
```
    4  0  4
 -  3  2  7
          7  7
```

7 How long is each line?

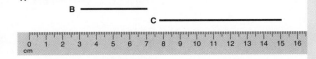

(a) Line A is about ___10___ cm long.

(b) Line B is about ___4___ cm long.

(c) Line C is about ___7___ cm long.

8 A pine tree is 147 feet tall.
A redwood tree is 178 feet taller than the pine tree.
How tall is the redwood tree?

147 + 178 = 325

The redwood tree is ___325___ ft tall.

9 A melon weighs 910 g.
A mango is 797 g lighter than the melon.
What does the mango weigh?

910 − 797 = 113

The mango weighs ___113___ g.

Challenge

10 There are four different weights: 1 lb, 2 lb, 3 lb, and 4 lb.
How many different weights can you get using only these weights?

Students should be able to see they can make 1 lb, 2 lb, 3 lb, and 4 lb with one weight only, and a maximum of 10 lb with all four weights. They can then test a few combinations of 2 or 3 weights to show they can get 5 lb, 6 lb, 7 lb, 8 lb, and 9 lb as well.

You can get ___10___ different weights.

Suggested number of class periods: 7–8

	Lesson	Page	Resources		Objectives
	Chapter Opener	p. 163	TB:	p. 129	Investigate operations on equal groups.
1	Multiplication — Part 1	p. 164	TB: WB:	p. 130 p. 123	Understand multiplication as an operation to find the total amount when we have equal groups.
2	Multiplication — Part 2	p. 166	TB: WB:	p. 133 p. 127	Understand that multiplication is commutative.
3	Practice A	p. 168	TB: WB:	p. 135 p. 131	Practice multiplication.
4	Division — Part 1	p. 170	TB: WB:	p. 138 p. 135	Understand division as sharing. Use division to find the size of one group given the total and the number of groups.
5	Division — Part 2	p. 173	TB: WB:	p. 142 p. 139	Understand division as grouping. Use division to find the number of groups when given the total and the number in each group.
6	Multiplication and Division	p. 176	TB: WB:	p. 146 p. 143	Use an array model to explore the relationship between multiplication and division.
7	Practice B	p. 178	TB: WB:	p. 149 p. 145	Practice multiplication and division.
	Workbook Solutions	p. 180			

In **Dimensions Math® 1B**, students learned to:

*   Recognize equal groups.
*   Find the total number in the groups by repeated addition.
*   Understand situations involving sharing a number of objects equally.
*   Understand situations involving making equal groups.

In this chapter, students formalize their knowledge of multiplication and division. The emphasis is on understanding multiplication and division using equal groups and arrays, not on memorizing facts. Students should be allowed to use counters or draw pictures as needed throughout the chapter.

Students begin by learning how to use the multiplication symbol to represent the operation of adding quantities in equal groups. They will learn that they can use a multiplication symbol to express (the number in each group) × (the number of groups), and will read this symbol as "times." 4 × 3 means 4 groups of 3, or 4 times 3.

In the next lesson, students will learn that the total is the same whether there are 4 groups of 3 or 3 groups of 4, that is, 4 × 3 = 3 × 4. The order of the factors when students first learn to write the multiplication expression is arbitrary. One could just as easily introduce multiplication as 4 × 3 = 4 multiplied by 3, that is, 4 in 3 groups. However, it is important that students start with an established meaning for each factor, so they understand that the product is the same for 4 groups of 3 or 3 groups of 4. That is, they need to understand that 4 × 3 = 3 × 4 when the first factor is the number of groups in both expressions, not because 4 groups of 3 is the same as 3 multiplied by 4 (which is also 4 groups of 3).

Later, students can write the factors in any order and will not be required to write the number of groups as the first factor in the expression. They will learn that there is no way to know whether the first factor in an expression represents the number of groups or the number in each group.

Sometimes the number in each group is called the multiplicand, and the number of groups is called the multiplier, and that multiplicand × multiplier = product. This order is the opposite of how students are writing expressions in this chapter.

Lesson 4 introduces division as sharing (partitive division). Sharing division begins with the total number of objects and puts them into a number of groups to find out how many there are in each group:

*   16 jelly beans are shared between 2 friends. How many jelly beans does each friend receive?
    16 ÷ 2 = 8

Lesson 5 introduces division as grouping (quotative division).

Division as grouping begins with the total number of objects and puts them into equal groups of a given amount to find out how many groups we can make.

*   There are 16 jelly beans. Each friend receives 2 jelly beans. How many friends will get jelly beans?
    16 ÷ 2 = 8

Students will be using the terms "sharing" and "grouping," not "partitive division" and "quotative division."

In **Dimensions Math® 3**, students will be more formally introduced to division with remainders. While this defies the "equal groups" definition of division, understanding remainders as fractions will be taught in **Dimensions Math® 4** after the concept of fractions as division is covered.

Materials

- 12 items to be shared
- Counters
- Craft sticks (optional)
- Paper, at least 12 cm long
- Small plates
- Whiteboards

Blackline Masters

- Score Sheet

Storybooks

- *How Do You Count a Dozen Ducklings?* by Seon Chae
- *Two Ways to Count to Ten: A Liberian Folktale* by Ruby Dee
- *Counting Sheep* by Julie Glass
- *Two of Everything* by Lily Toy Hong
- *The Doorbell Rang* by Pat Hutchins
- *Spunky Monkeys on Parade* by Stuart J. Murphy
- *Corkscrew Counts: A Story About Multiplication* by Donna Jo Napoli and Richard Tchen
- *A Remainder of One* by Elinor J. Pinczes
- *Arctic Fives Arrive* by Elinor J. Pinczes

Activities

Games and activities included in this chapter are designed to provide practice for basic multiplication and division. They can be used after students complete the **Do** questions, or anytime review and practice are needed.

Notes

Chapter Opener

Objective

- Investigate operations on equal groups.

Materials

- Counters
- Small plates

Provide pairs of students with counters and paper plates. Have them represent the potatoes, avocados, eggplant, and squash and find a way to find how many vegetables of each kind there are without counting each vegetable.

Discuss student strategies to find the total.

Activity

▲ Equal Group Exploration

Materials: 40 counters, small plates

This partner activity encourages students to think about different examples of equal groups.

Partner 1 chooses some plates and some counters and puts an equal number of counters, up to 10, on each plate. Any extra counters can be returned to the pile.

Partner 2 counts the counters and says or writes an equation:

For example:

- There are 4 threes, that's 12.
- 4 groups of 3 makes 12.
- 3 + 3 + 3 + 3 = 12

Then, partners can trade jobs and repeat.

Chapter 6

Multiplication and Division

How many of each kind of vegetable are there?

How can we find how many there are quickly without counting each one?

ARITHMETICVILLE
FARMER'S MARKET
FRESH FRUITS, VEGETABLES & MORE!

129

129

Lesson 1 Multiplication — Part 1

Objective

- Understand multiplication as an operation to find the total amount when we have equal groups.

Materials

- Counters

Think

Provide students with counters and pose the **Think** problem with the hair ties.

Have students write an equation and share how they found their answers. Discuss the different representations students may have with counters.

They may have:

- 4 piles of 3 counters
- 4 rows of 3 counters

Equations could be:

- 3 + 3 + 3 + 3 = 12
- 4 + 4 + 4 = 12

Learn

Introduce the "×" as the symbol for multiplication. Have students discuss the different examples of multiplication. Emma uses repeated addition. Dion uses multiplication.

Ask students if they find it tedious to write out long addition equations. The multiplication symbol is a shorter way of writing an equation when the groups are equal.

Remind students that when we multiply, we always have equal groups.

Have students compare their methods from **Think** to the one shown in the textbook. Provide students with additional problems and counters as needed before moving on to **Do**.

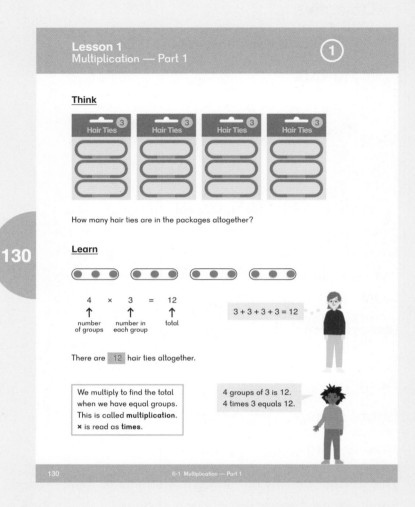

Do

Have students show the numbers with counters only if needed. Most should be working with just the pages in the textbook.

3–**5** Encourage students to use whiteboards to draw the problem.

Activity

▲ **Multiplication Mania**

Materials: 2 dice, counters, dot paper

Player 1 rolls the dice and chooses one of the numbers as the number of groups, the other number is the number of counters in each group.

For example, if Player 1 rolls:

They could call one of the following:

- 5 groups of 2
- 5 × 2
- 2 groups of 5
- 2 × 5

★ Player 2 makes groups with the counters and finds the total.

Instead of using counters, have Player 2 circle the equal groups on dot paper.

Exercise 1 • page 123

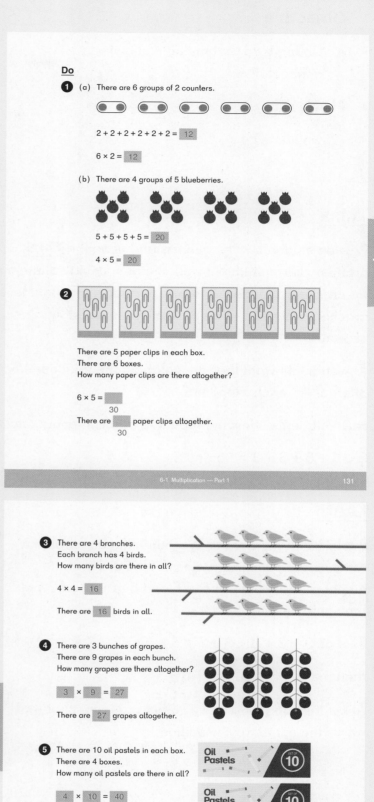

Objective

- Understand that multiplication is commutative.

Materials

- Counters

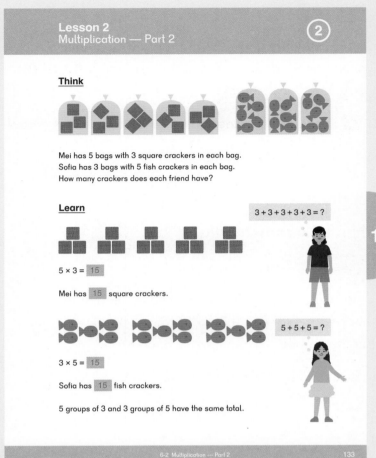

Think

Mei has 5 bags with 3 square crackers in each bag.
Sofia has 3 bags with 5 fish crackers in each bag.
How many crackers does each friend have?

Learn

$3 + 3 + 3 + 3 + 3 = ?$

$5 \times 3 = \boxed{15}$

Mei has $\boxed{15}$ square crackers.

$5 + 5 + 5 = ?$

$3 \times 5 = \boxed{15}$

Sofia has $\boxed{15}$ fish crackers.

5 groups of 3 and 3 groups of 5 have the same total.

6-2 Multiplication — Part 2   133

Think

Provide students with counters and pose the **Think** problem. Have students work in pairs, using counters for crackers. Suggest that one partner represent Mei's crackers while the other partner represents Sofia's crackers.

Ask students what they notice about the total number of crackers each friend has.

Have students show the repeated addition equations:

$3 + 3 + 3 + 3 + 3 = 15$ and $5 + 5 + 5 = 15$

Learn

Explain that the answer to a multiplication problem remains the same regardless of the order of the numbers. Both 3 groups of 5 and 5 groups of 3 yield the same answer.

Just as $3 + 3 + 3 + 3 + 3 = 5 + 5 + 5$,

then $5 \times 3 = 3 \times 5$.

Ask if they can think of any other math concept that works the same way. (Addition)

Teacher's Guide 2A Chapter 6

Do

These problems emphasize that it does not matter which is the number of groups and which is the number in each group. In both situations, the answer is the same.

Extend by posing problems with large numbers that students may not be able to calculate, such as:

$100 \times 2 = 2 \times ?$ What is the missing number?

❷ 8 rows of 3 is the same as 3 columns of 8.
8 groups of 3 is the same as 3 groups of 8.

Activity

▲ Array Punch

Materials: Small rectangular sheets of construction paper, hole punches, glue

Have each student fold a piece of paper to create 8 work spaces.

Give students 4 multiplication equations.

Students will punch out an array for their first equation and glue it onto the left side of the paper. They can use this array to solve the problem.

Then, they use the circles that were punched out to build another array with rows and columns. Have them show a different array from their punched-out array.

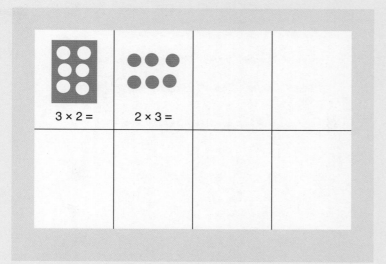

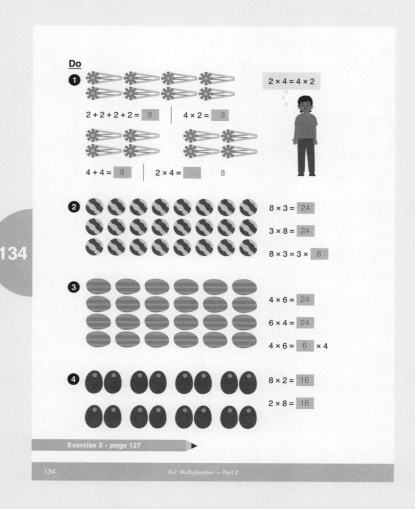

134

Exercise 2 • page 127

Lesson 3 Practice A

Objective

- Practice multiplication.

Practice

Students will continue to practice multiplication in future lessons in **Dimensions Math® 2A** and **2B**.

2 — **5** Students won't be formally introduced to the term "array" until **Lesson 6**. These problems (and the included activities) provide the opportunity to explore and discuss the layout of rows and columns.

Students should note that it is easy to count the groups in an array format in either order.

Students who struggle with **3** — **5** can draw arrays on whiteboards and circle the quantities in a row or column to find the numbers to be multiplied. They should also play any of the games from the chapter before tackling the word problems in **6**.

6 Students have not yet multiplied measurements. Since the objects being measured are not distinct quantities, some students may have difficulty with these problems. However, students have learned to think of measurement in terms of units. For (c), they can envision a balance with five 1-kg weights balancing a brick. Therefore, there are 2 groups of 5 kg (five 1-kg units). Similarly, for (d), there are 10 groups of $3 (three $1 units), and for (e), there are 7 groups of 2 miles (two 1-mile units).

Although it was not introduced in the chapter on length, many students will be able to think of a mile as a unit of measurement as they have probably been exposed to the idea in real life.

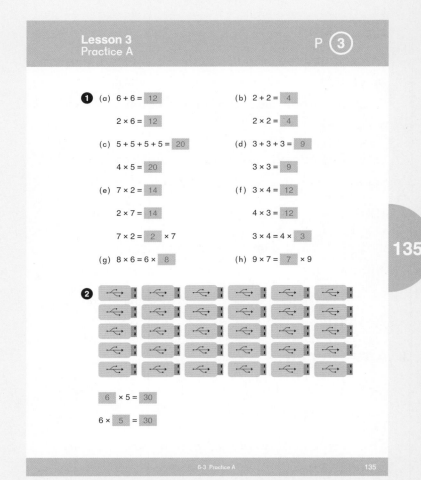

Teacher's Guide 2A Chapter 6   © 2017 Singapore Math Inc.

Activities

▲ Fences

Materials: Centimeter or half-inch graph paper in dry erase sleeves, 2 dice, 2 colors of dry erase markers

Students take turns rolling the dice. They fence in land on the game board by outlining an array with the same number of spaces as the dice roll.

The players write two multiplication equations on their newly acquired land.

In the example shown below:

- Player 1 (blue) rolled 5 and 4 and fenced in a 5 × 4 array.
- Player 2 (red) rolled 2 and 6 and fenced in a 2 × 6 array.

Play stops when a player can't fit their array on the board. Each player adds up the total amount of land (or boxes) they have fenced in. The player with the most land is the winner.

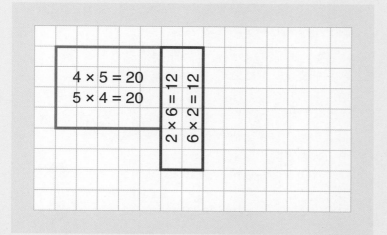

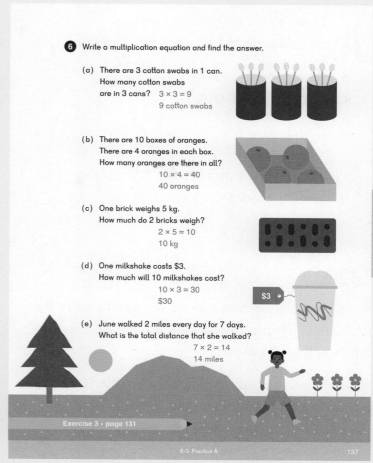

6 Write a multiplication equation and find the answer.

(a) There are 3 cotton swabs in 1 can. How many cotton swabs are in 3 cans? 3 × 3 = 9
9 cotton swabs

(b) There are 10 boxes of oranges. There are 4 oranges in each box. How many oranges are there in all?
10 × 4 = 40
40 oranges

(c) One brick weighs 5 kg. How much do 2 bricks weigh?
2 × 5 = 10
10 kg

(d) One milkshake costs $3. How much will 10 milkshakes cost?
10 × 3 = 30
$30

(e) June walked 2 miles every day for 7 days. What is the total distance that she walked?
7 × 2 = 14
14 miles

Exercise 3 · page 131

6-3 Practice A

137

▲ Multiplication Stories

Materials: Markers or crayons, art paper

Assign or have students select an equation to illustrate. Have them write the problem and equation and then illustrate.

Examples:

- There are 3 dogs. Each dog has 3 diamonds on his collar. How many diamonds in all?
- 2 friends each had 6 bracelets on one of their arms. How many bracelets were there?
- The sports shop had 5 basketballs, 5 footballs, and 5 soccer balls. How many balls did they have in all?

Exercise 3 · page 131

Objectives

- Understand division as sharing.
- Use division to find the size of one group given the total and the number of groups.

Lesson Materials

- Counters
- Small plates
- 12 items to share (cookies, crackers, trading cards)

Think

Have students work in groups of three to solve the problem in **Think**. Provide the groups with counters and small paper plates.

Have groups act out and discuss solutions. Students will get the correct quotient of 4, however, they may not be able to explain how they solved the problem. Have 3 students come to the front of the room. Share the 12 objects between students one at a time so that they get an equal number.

Reinforce the concept of equal groups. When we multiply, we are finding a total. When we divide, we start with a total and (in this case) are finding the number in each group.

Remind students that when we add or subtract, the groups do not have to be equal, but when we multiply and divide there will be equal groups.

Learn

Discuss the problem with the cookies as shown in the textbook. Introduce the term "division" and the division sign.

Note: This lesson provides examples of division as sharing (partitive division). The next lesson will focus on division as grouping.

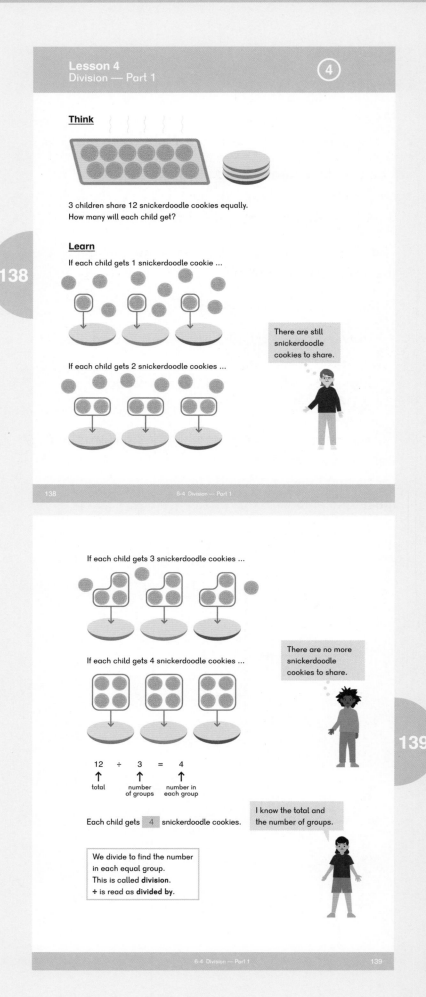

Provide students with additional examples of "how many in a group" using counters. For example:

- 20 cookies shared equally among 4 children.
- 16 dollar bills divided equally among 4 children.
- 30 pencils shared equally between 5 children.

Have students create similar situations to describe sharing before working through the **Do** portion. Students may come up with problems with larger numbers. They do not need to find an answer. Examples:

- 100 cheeseburgers shared equally among 5 friends.
- A million dollars shared equally between me and my brother.

Do

Students should be able to complete these problems without counters. Note the visual progression of the problems: ❶ — ❸ show grouped items.

❹ Unlike the previous problems, the hot peppers are not divided into groups.

❺ — ❻ Students may need to use counters or draw pictures to solve the problems.

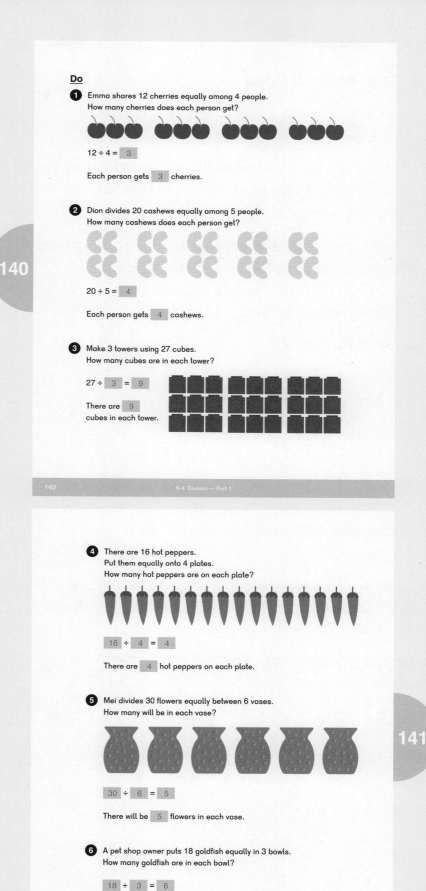

Activities

▲ The Doorbell Rang ...

Materials: *The Doorbell Rang* by Pat Hutchins, counters

This is a great time to read *The Doorbell Rang* by Pat Hutchins and have students act out the story with their counters.

In the book, two children begin by sharing 12 cookies between the two of them. Each time the doorbell rings, more friends arrive and the cookies are divided among all the children.

▲ Student Sorts

Have students put themselves into groups before lining up for recess or another class.

- Line up in 2 rows.
- Put yourselves into 5 groups. How many in a group? How many in all?

Try blowing a whistle a certain number of times and have students get into groups of that number and link arms.

Students without a group get to help the teacher decide the next number given to the other students.

Exercise 4 • page 135 ▶

Lesson 5 Division — Part 2

Objectives

- Understand division as grouping.
- Use division to find the number of groups when given the total and the number in each group.

Materials

- Counters
- Craft sticks (optional)

Think

Have students work the lychee problem in **Think**. They can use counters to act out the problem in small groups.

Ask students how this problem is similar to the problem from the previous lesson. How is it different?

Students should note that the numbers are the same and they are still dividing.

They should also note that in the previous lesson, they knew how many children received cookies, but didn't know how many cookies each child received. That is, they know the **number** of groups, but not **how many** in each group.

In this example, students know how many lychees are in each bowl, but they don't know how many bowls they need. That is, they know **how many** are in each group, but not the **number** of groups.

Learn

Discuss the problem with the lychees as shown in the textbook. Have students compare the division equation from the previous lesson.

Provide students with additional examples of "how many groups" using counters:

- 20 cookies, each child receives 4.
- 16 dollar bills, each child receives 4.
- 30 pencils, each child receives 5.

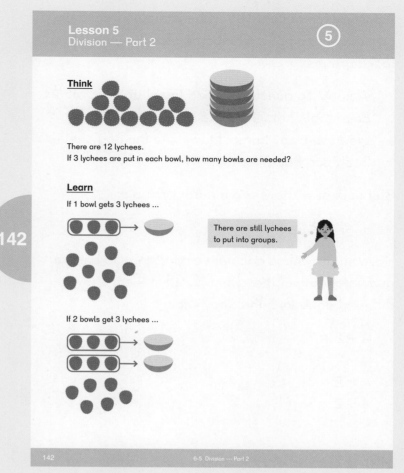

Think

There are 12 lychees.
If 3 lychees are put in each bowl, how many bowls are needed?

Learn

If 1 bowl gets 3 lychees ...

There are still lychees to put into groups.

If 2 bowls get 3 lychees ...

142   6-5 Division — Part 2

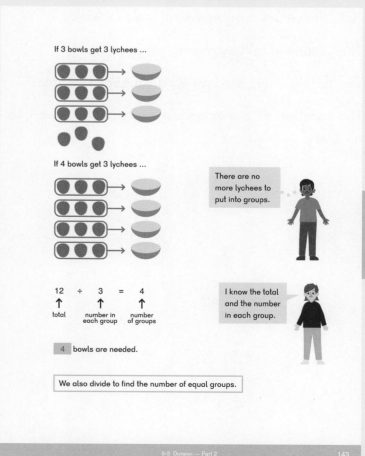

If 3 bowls get 3 lychees ...

If 4 bowls get 3 lychees ...

There are no more lychees to put into groups.

143

$$12 \div 3 = 4$$

↑ total   ↑ number in each group   ↑ number of groups

I know the total and the number in each group.

4 bowls are needed.

We also divide to find the number of equal groups.

6-5 Division — Part 2   143

Ask students to come up with examples from their own experience of a situation where they had to make groups.

- We had 16 guests and had to set up card tables. Each table held 4 people.
- We gave away bags of towels to the dog shelter. Each bag had 3 towels.

Students now know seven math symbols:

$+$, $-$, $=$, $>$, $<$, $\times$, and $\div$

Have students use counters or drawings to ensure understanding of the symbols. Have them show different equations that they know:

- $4 + 2 = ?$
- $4 - 2 = ?$
- $4 \times 2 = ?$
- $4 \div 2 = ?$

Do

Note the visual progression of the problems.

1 The groups of buttons are circled.

2 The tortilla chips are loosely arranged in groups.

3 The cranberries are arranged in organized towers (or an array).

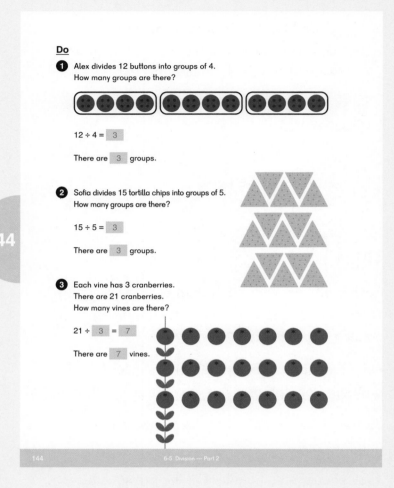

144

Do

1 Alex divides 12 buttons into groups of 4. How many groups are there?

$12 \div 4 = \boxed{3}$

There are $\boxed{3}$ groups.

2 Sofia divides 15 tortilla chips into groups of 5. How many groups are there?

$15 \div 5 = \boxed{3}$

There are $\boxed{3}$ groups.

3 Each vine has 3 cranberries. There are 21 cranberries. How many vines are there?

$21 \div \boxed{3} = \boxed{7}$

There are $\boxed{7}$ vines.

144   6-5 Division — Part 2

④ Unlike the previous problems, the crayons are not divided into groups.

⑤—⑥ Students may need counters, craft sticks, or to draw pictures to solve the problems.

Activities

▲ Student Sorts

Extend **Student Sorts** from the previous lesson to include both types of division:

- Line up in three rows.
- Sort yourselves into groups of 3. How many groups are there?
- How many students are left over?

▲ Division Stories

Assign or have students select an equation to illustrate. Have them write the problem and equation and then illustrate. Examples:

- I counted 20 dog legs in the park. How many dogs are in the park?
- 6 friends shared 12 bracelets equally. How many bracelets did each friend have?
- The sports shop had 18 balls. They put them into bags of 3. How many bags did the shop make?

$20 \div 4 = ?$

Exercise 5 · page 139

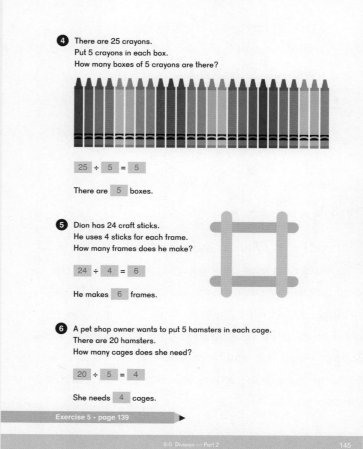

④ There are 25 crayons.
Put 5 crayons in each box.
How many boxes of 5 crayons are there?

$25 \div 5 = 5$

There are 5 boxes.

⑤ Dion has 24 craft sticks.
He uses 4 sticks for each frame.
How many frames does he make?

$24 \div 4 = 6$

He makes 6 frames.

⑥ A pet shop owner wants to put 5 hamsters in each cage.
There are 20 hamsters.
How many cages does she need?

$20 \div 5 = 4$

She needs 4 cages.

Exercise 5 · page 139

6-5 Division — Part 2   145

145

Lesson 6 Multiplication and Division

Objective

- Use an array model to explore the relationship between multiplication and division.

Lesson Materials

- Counters

Think

In this **Think**, Dion introduces the term "array." Students can use counters to create an array and circle the rows and columns to find different multiplication and division equations for the basketballs. At this point, some students may no longer need the counters and should be encouraged to simply draw the array or write the equations on a whiteboard.

From the picture of the basketball cart, ask students:

- Where do you see 3 groups of 5?
- Where do you see 5 groups of 3?
- Divide 15 into 3 equal groups. How many are in each group?
- Divide 15 into groups of 3. How many groups are there?

Learn

Have students discuss the grouping in the textbook. Ask them what other math situations they know that have four related equations.

Do

Students should be able to complete these problems without counters.

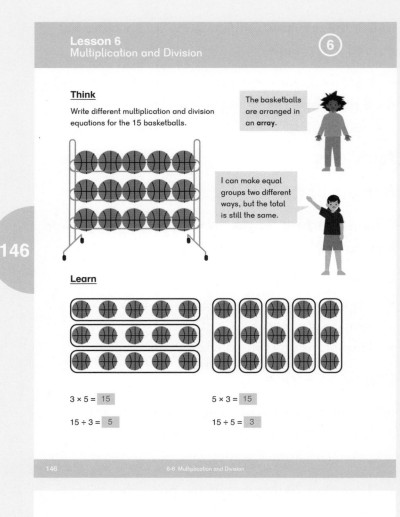

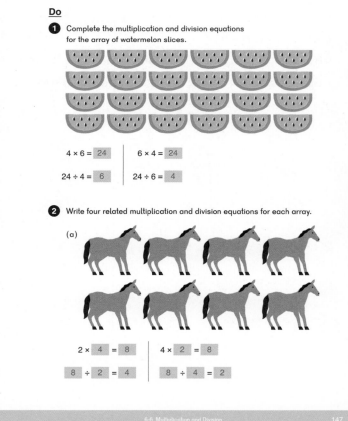

2 (c) Discuss Sofia's thought about the paper planes.

3 Continue to practice this skill by providing counters or other manipulatives. Have one partner make an array and the other partner find the related multiplication and division sentences.

Students can play the **Fences** game from **Lesson 3: Practice A** or **Hip! Hip! Array!** below for further practice.

★ To extend, ask students questions in which numbers are missing from an array. For example:

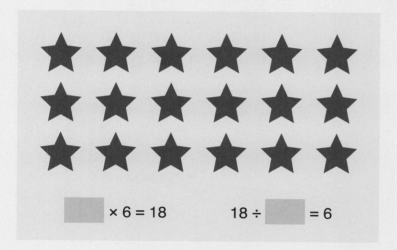

$$\boxed{} \times 6 = 18 \qquad 18 \div \boxed{} = 6$$

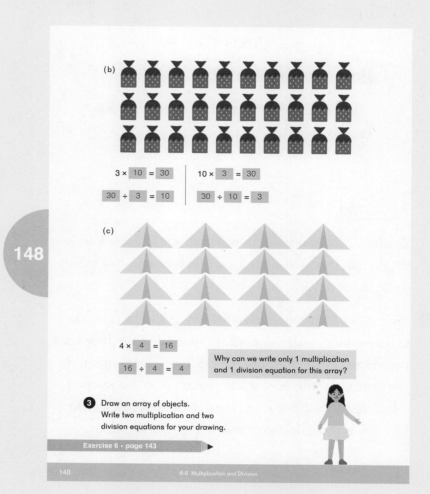

(b)

$$3 \times \boxed{10} = 30 \qquad \Big| \qquad 10 \times \boxed{3} = 30$$

$$\boxed{30} \div \boxed{3} = \boxed{10} \qquad \boxed{30} \div \boxed{10} = \boxed{3}$$

(c)

$$4 \times \boxed{4} = \boxed{16}$$

$$\boxed{16} \div \boxed{4} = \boxed{4}$$

Why can we write only 1 multiplication and 1 division equation for this array?

3 Draw an array of objects. Write two multiplication and two division equations for your drawing.

Exercise 6 • page 143

Activity

▲ Hip Hip Array

Materials: 2 dice, Score Sheet (BLM) for each player

In each round, players take turns rolling the dice and drawing an array.

The player records the multiplication equation derived from the array on a score sheet and solves it.

Players compare their answers. The player with the greatest product wins the round and puts a check by her equation. The player with the most wins after 9 rounds is the overall winner.

Exercise 6 • page 143

Objective

- Practice multiplication and division.

Lesson Materials

- Paper, at least 12 cm long to be cut into strips

Practice

Students will continue to work with multiplication and division in the next chapter while learning the tables for 2, 5, and 10.

3 This is the first time students are asked to consider division with measurement. If needed, have them cut two paper strips that are 12 cm long. They can cut the first strip into two equal pieces and the second into pieces that are 2 cm long. Note the answer to (a) is 6 cm, while the answer to (b) is 6 pieces.

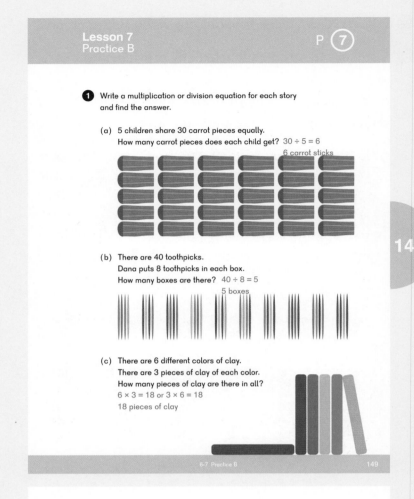

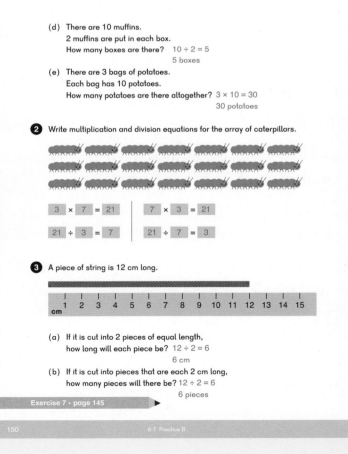

Brain Works

★ **Math Libs**

Have students create word problems and write equations with numbers that they know.

> There were _____ _____.
>   (number)   (animals)
>
> Each of them had _____ _____.
>   (another number) (objects)
>
> How many _____ were there in all?
>   (same objects)
>
> -
>
> _____ had _____ _____.
> (name)   (number)   (objects)
>
> _____ put them into _____ bags.
> (same name)   (another number)
>
> How many _____ were in each bag?
>   (same objects)
>
> -
>
> _____ had _____ _____.
> (name)   (number)   (objects)
>
> _____ put _____ into each box.
> (same name) (another number)
>
> How many boxes did _____ need?
>   (same name)

Exercise 7 • page 145 ▶

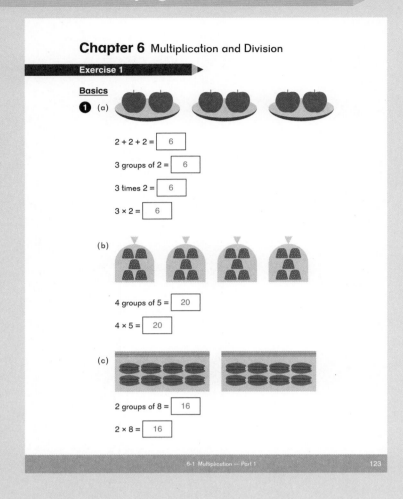

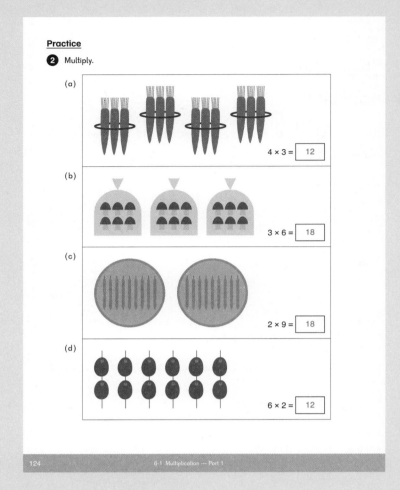

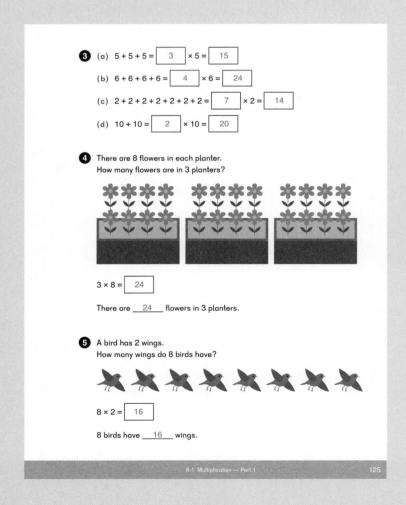

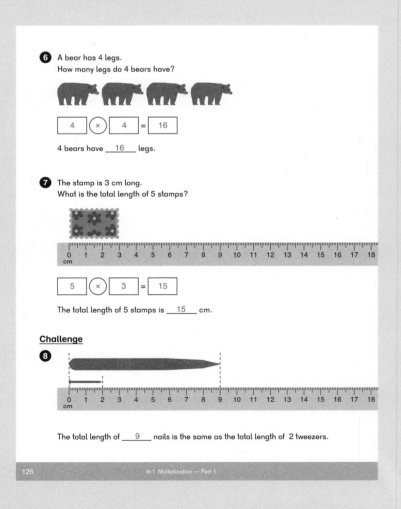

Teacher's Guide 2A Chapter 6 © 2017 Singapore Math Inc.

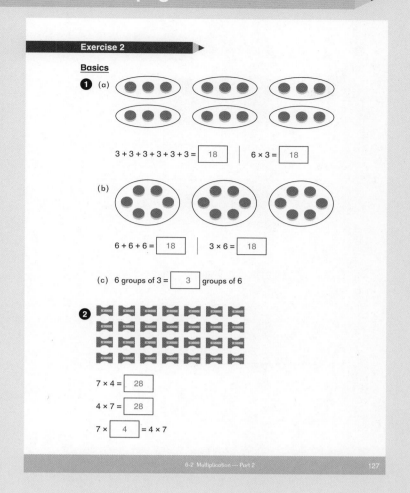

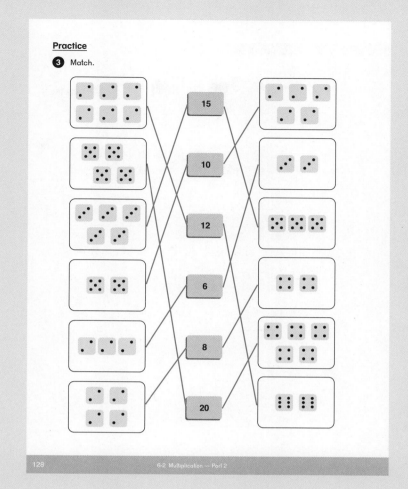

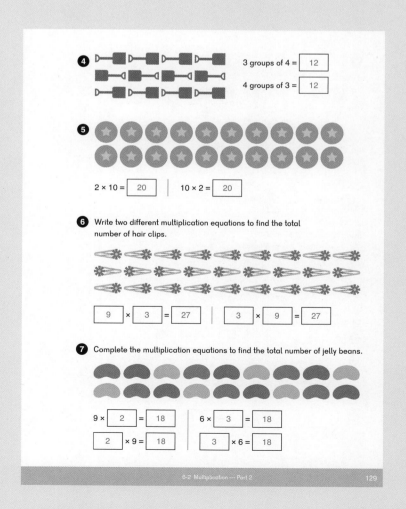

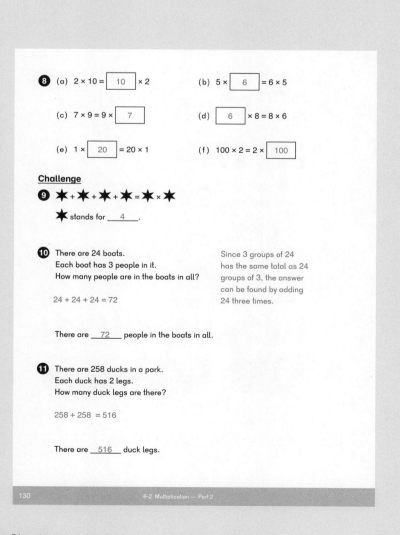

Exercise 3

Check

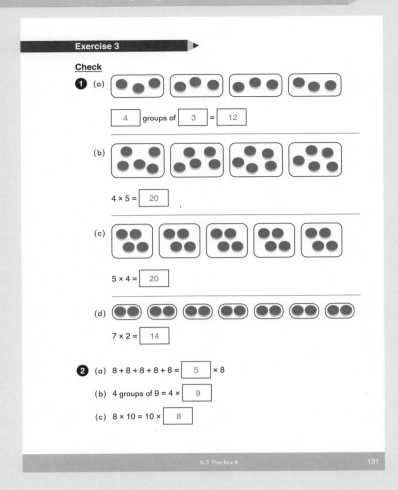

1 (a) [4] groups of [3] = [12]

(b) 4 × 5 = [20]

(c) 5 × 4 = [20]

(d) 7 × 2 = [14]

2 (a) 8 + 8 + 8 + 8 + 8 = [5] × 8

(b) 4 groups of 9 = 4 × [9]

(c) 8 × 10 = 10 × [8]

3 Draw a picture to show each of the following:

(a) 3 groups of 3

(b) 8 groups of 2

(c) 2 × 8

(d) 4 × 10

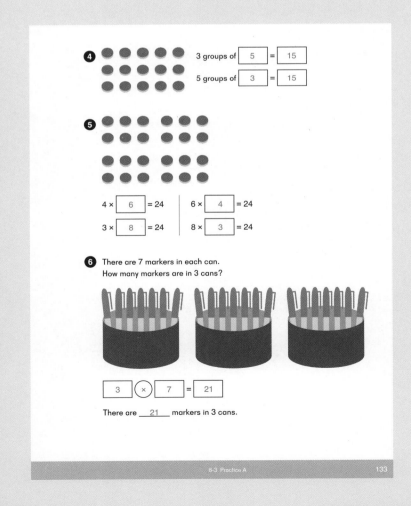

4 3 groups of [5] = [15]

5 groups of [3] = [15]

5 4 × [6] = 24   6 × [4] = 24

3 × [8] = 24   8 × [3] = 24

6 There are 7 markers in each can.
How many markers are in 3 cans?

[3] ⓧ [7] = [21]

There are __21__ markers in 3 cans.

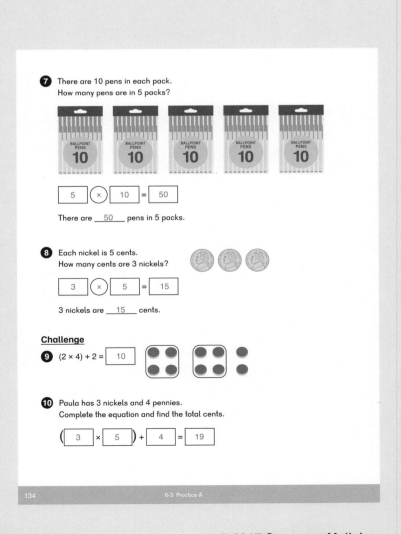

7 There are 10 pens in each pack.
How many pens are in 5 packs?

[5] ⓧ [10] = [50]

There are __50__ pens in 5 packs.

8 Each nickel is 5 cents.
How many cents are 3 nickels?

[3] ⓧ [5] = [15]

3 nickels are __15__ cents.

Challenge

9 (2 × 4) + 2 = [10]

10 Paula has 3 nickels and 4 pennies.
Complete the equation and find the total cents.

([3] × [5]) + [4] = [19]

Exercise 4

Basics

1 There are 18 balls.
3 people share them equally.
How many does each person get?
Draw more balls to show 18 balls shared into 3 equal groups.

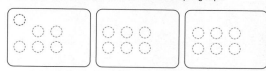

Each person gets __6__ balls.

18 shared into __3__ groups is __6__ in each group.

18 ÷ 3 = [6]

18 divided by __3__ equals __6__.

2 Draw more cherries to show 15 cherries divided into 5 equal groups.

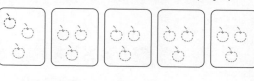

There are __3__ cherries in each group.

15 ÷ 5 = [3]

Practice

3 Aki put 21 markers equally into 3 cans.

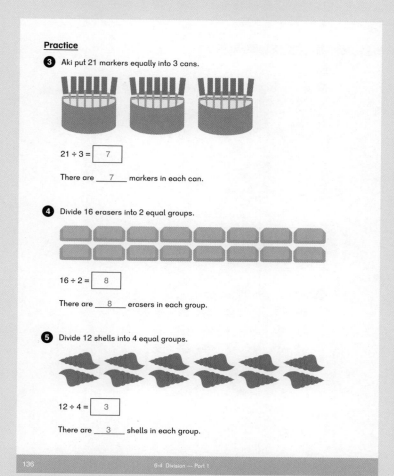

21 ÷ 3 = [7]

There are __7__ markers in each can.

4 Divide 16 erasers into 2 equal groups.

16 ÷ 2 = [8]

There are __8__ erasers in each group.

5 Divide 12 shells into 4 equal groups.

12 ÷ 4 = [3]

There are __3__ shells in each group.

6 Draw more ✗ to find the number in each group.

(a) 24 divided into 6 groups

24 ÷ 6 = [4]

(b) 24 divided into 4 groups

24 ÷ 4 = [6]

(c) 15 divided into 3 groups

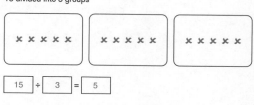

15 ÷ 3 = [5]

7 Jordan put 24 beads equally on 3 strings.
Draw more beads to find how many are on each string.

24 ÷ 3 = 8

There are __8__ beads on each string.

Challenge

8 Sebastian wants to share 9 crayons equally with a friend.

Can he share them equally? __no__

How many crayons are left over? __1__

9 Divide 17 counters into 5 equal groups, with as few left over as possible.
How many are left over?

17 ÷ 5 is __3__ with __2__ left over.

© 2017 Singapore Math Inc.   Teacher's Guide 2A Chapter 6   183

Exercise 5

Basics

1 Mayam has 18 balls.
She put 3 balls in each can.
How many cans does she use?

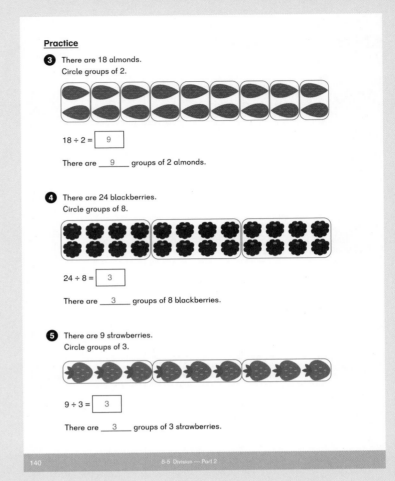

____6____ cans are used.

18 grouped by ___3___ is ___6___ groups.

$18 \div 3 = \boxed{6}$

18 divided by ___3___ equals ___6___.

2 Divide 35 beads into groups of 7.

There are ___5___ groups.

$35 \div 7 = \boxed{5}$

Practice

3 There are 18 almonds.
Circle groups of 2.

$18 \div 2 = \boxed{9}$

There are ___9___ groups of 2 almonds.

4 There are 24 blackberries.
Circle groups of 8.

$24 \div 8 = \boxed{3}$

There are ___3___ groups of 8 blackberries.

5 There are 9 strawberries.
Circle groups of 3.

$9 \div 3 = \boxed{3}$

There are ___3___ groups of 3 strawberries.

6 Find the number of groups.

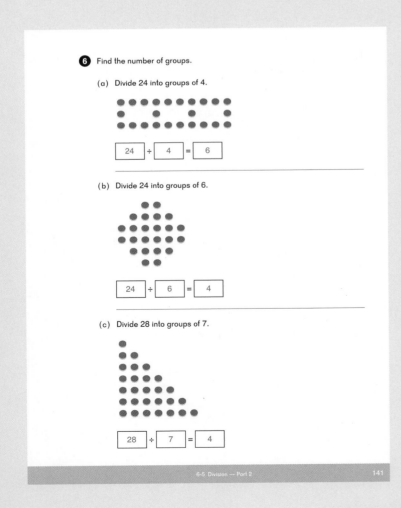

(a) Divide 24 into groups of 4.

$\boxed{24} \div \boxed{4} = \boxed{6}$

(b) Divide 24 into groups of 6.

$\boxed{24} \div \boxed{6} = \boxed{4}$

(c) Divide 28 into groups of 7.

$\boxed{28} \div \boxed{7} = \boxed{4}$

7 Luke has 18 eggs.
He puts 6 eggs in each carton.
How many egg cartons does he need?

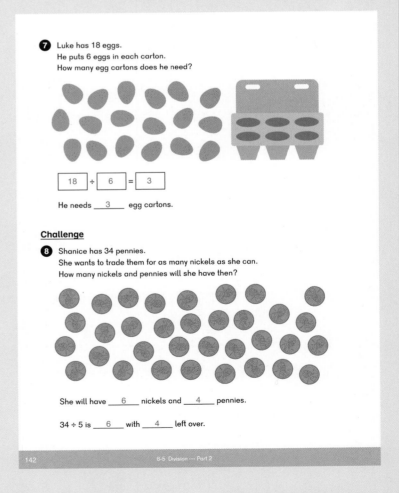

$\boxed{18} \div \boxed{6} = \boxed{3}$

He needs ___3___ egg cartons.

Challenge

8 Shanice has 34 pennies.
She wants to trade them for as many nickels as she can.
How many nickels and pennies will she have then?

She will have ___6___ nickels and ___4___ pennies.

$34 \div 5$ is ___6___ with ___4___ left over.

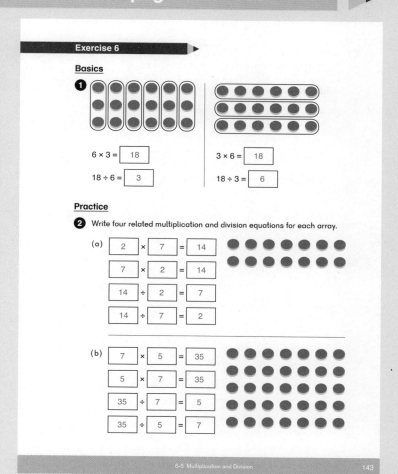

Exercise 6

Basics

1

$6 \times 3 = \boxed{18}$

$18 \div 6 = \boxed{3}$

$3 \times 6 = \boxed{18}$

$18 \div 3 = \boxed{6}$

Practice

2 Write four related multiplication and division equations for each array.

(a)
$\boxed{2} \times \boxed{7} = \boxed{14}$

$\boxed{7} \times \boxed{2} = \boxed{14}$

$\boxed{14} \div \boxed{2} = \boxed{7}$

$\boxed{14} \div \boxed{7} = \boxed{2}$

(b)
$\boxed{7} \times \boxed{5} = \boxed{35}$

$\boxed{5} \times \boxed{7} = \boxed{35}$

$\boxed{35} \div \boxed{7} = \boxed{5}$

$\boxed{35} \div \boxed{5} = \boxed{7}$

6-6 Multiplication and Division   143

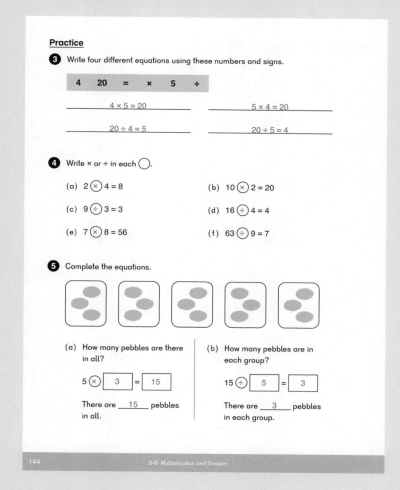

Practice

3 Write four different equations using these numbers and signs.

| 4 | 20 | = | × | 5 | ÷ |

$4 \times 5 = 20$ $5 \times 4 = 20$

$20 \div 4 = 5$ $20 \div 5 = 4$

4 Write × or ÷ in each ◯.

(a) $2 \, ⓧ \, 4 = 8$

(b) $10 \, ⓧ \, 2 = 20$

(c) $9 \, ⊙ \, 3 = 3$

(d) $16 \, ⊙ \, 4 = 4$

(e) $7 \, ⓧ \, 8 = 56$

(f) $63 \, ⊙ \, 9 = 7$

5 Complete the equations.

(a) How many pebbles are there in all?

$5 \, ⓧ \, \boxed{3} = \boxed{15}$

There are ___15___ pebbles in all.

(b) How many pebbles are in each group?

$15 \, ⊙ \, \boxed{5} = \boxed{3}$

There are ___3___ pebbles in each group.

144   6-6 Multiplication and Division

Exercise 7

Check

1 (a) $5 + 5 + 5 + 5 = \boxed{4} \times 5 = \boxed{20}$

(b) $3 + 3 + 3 = \boxed{3} \times 3 = \boxed{9}$

(c) $6 + 6 = \boxed{2} \times 6 = \boxed{12}$

(d) $10 + 10 + 10 + 10 + 10 + 10 + 10 + 10 = \boxed{8} \times 10 = \boxed{80}$

2 Draw 3 flowers in each vase.
How many flowers are there in all?

$6 \times 3 = \boxed{18}$

There are ___18___ flowers in all.

3 Draw another 3 tanks with 4 fish in each tank.
How many fish are there in all?

$4 \times 4 = \boxed{16}$

There are ___16___ fish in all.

4 A ribbon is 12 cm long.

(a) If it is cut into pieces each 6 cm long, how many pieces will there be?

$12 \div 6 = \boxed{2}$

There will be ___2___ pieces.

(b) If it is cut in 3 pieces each the same length, how long will each piece be?

$12 \div 3 = \boxed{4}$

Each piece will be ___4___ cm long.

5 There are 30 cookies.

(a) If the cookies are put equally into 5 boxes, how many cookies will be in each box?

$30 \div 5 = \boxed{6}$

There will be ___6___ cookies in each box.

(b) If 10 cookies are put in each box, how many boxes will there be?

$30 \div 10 = \boxed{3}$

There will be ___3___ boxes.

6 Write two multiplication and two division equations using these numbers.

| 50 | 10 | 5 |

$\boxed{5} \times \boxed{10} = \boxed{50}$ $\boxed{50} \div \boxed{5} = \boxed{10}$

$\boxed{10} \times \boxed{5} = \boxed{50}$ $\boxed{50} \div \boxed{10} = \boxed{5}$

7

$\boxed{2} \times 7 = 14$ $\quad 7 \times \boxed{2} = 14$

$14 \div \boxed{2} = 7$ $\quad 14 \div 7 = \boxed{2}$

8 Each spider has 8 legs.

(a) If there are 3 spiders,
how many legs are there?

$3 \times \boxed{8} = \boxed{24}$

There are ___24___ legs.

(b) If there are 16 legs,
how many spiders are there?

$16 \div \boxed{8} = \boxed{2}$

There are ___2___ spiders.

9  $3 \times 4 = 6 \times \boxed{2}$

Challenge

10  $20 \div 2 = 5 \times \boxed{2}$

11 Write $+$, $-$, $\times$ or $\div$ in each $\bigcirc$.

(a) $10 \times 5 = 50$ (b) $10 \div 5 = 2$

(c) $10 - 5 = 5$ (d) $2 \times 10 = 20$

(e) $4 + 2 = 6$ (f) $8 \div 4 = 2$

(g) $8 \div 2 = 4$ (h) $8 - 2 = 6$

(i) $3 \times 3 = 9$ (j) $3 + 3 = 6$

(k) $6 - 3 = 3$ (l) $6 \div 3 = 2$

12 $6 \times 5 = 5 + 5 + 5 + 5 + \boxed{10}$

13 $7 \times 4 = 4 + 4 + 4 + 4 + \boxed{12}$

Suggested number of class periods: 12–13

	Lesson	Page	Resources		Objectives
	Chapter Opener	p. 191	TB:	p. 151	Investigate multiplication and division.
1	The Multiplication Table of 5	p. 192	TB: WB:	p. 152 p. 149	Build and understand the structure of the multiplication table of 5. Look for patterns in the multiplication table of 5.
2	Multiplication Facts of 5	p. 195	TB: WB:	p. 155 p. 151	Understand the commutative property in the multiplication table of 5. Learn the multiplication facts of 5.
3	Practice A	p. 198	TB: WB:	p. 158 p. 153	Practice the multiplication facts of 5.
4	The Multiplication Table of 2	p. 200	TB: WB:	p. 160 p. 157	Build and understand the structure of the multiplication table of 2. Look for patterns in the multiplication table of 2. Multiply numbers within the multiplication table of 2.
5	Multiplication Facts of 2	p. 203	TB: WB:	p. 163 p. 159	Understand the commutative property in the multiplication table of 2. Learn the multiplication facts of 2.
6	Practice B	p. 206	TB: WB:	p. 166 p. 161	Practice multiplication of 2 and 5.
7	The Multiplication Table of 10	p. 208	TB: WB:	p. 168 p. 165	Build and understand the structure of the multiplication table of 10. Look for patterns in the multiplication table of 10. Learn the multiplication facts of 10.
8	Dividing by 2	p. 212	TB: WB:	p. 172 p. 167	Use a related multiplication sentence to solve division problems without remainder where the divisor is 2.
9	Dividing by 5 and 10	p. 215	TB: WB:	p. 175 p. 171	Use a related multiplication sentence to solve division problems without remainder where the divisor is 5 or 10.
10	Practice C	p. 217	TB: WB:	p. 178 p. 175	Practice multiplication and division of 2, 5, and 10.
11	Word Problems	p. 218	TB: WB:	p. 179 p. 177	Solve problems involving multiplication and division. Determine the correct operation to solve a multiplication or division problem.

Lesson	Page	Resources	Objectives
Review 2	p. 221	TB: p. 182 WB: p. 181	Review content from Chapter 1 through Chapter 7.
Workbook Pages	p. 224		

In this chapter, students formalize their knowledge of multiplication and division for facts of 2, 5, and 10.

Lessons 1, 4, and 7 introduce building multiplication tables of 5, 2, and 10. Lessons 2, 5, and 7 have students practicing the facts for 5, 2, and then 10.

In Lessons 8 and 9, students learn to find their division facts by thinking of a related multiplication fact.

Most games and activities from this chapter can be adapted for other numbers in the multiplication and division tables. They can be used throughout the year and many will be included in **Dimensions Math®
2B Chapter 9: Multiplication and Division of
3 and 4.** Students should continue to practice these facts until they know them from memory.

Multiplication and division can be understood using the part-whole concept, however, as the relationship is multiplicative, number bonds are not used. A correct number bond for 2 × 4 = 8 is possible, however, when students move to problems such as 10 × 3 or 10 threes, number bonds are not helpful and bar models are used instead.

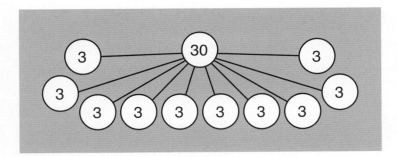

Bar Models

Lesson 11 makes use of multiplicative part-whole bar models. With multiplication, we know both the number of equal groups and the quantity in each group. The whole or total is missing.

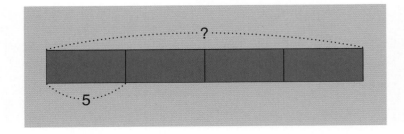

With division, the whole or total is known and we either know the number of equal groups or how many are in each group. In this example, we know there are 4 groups. We are looking for the number in each group.

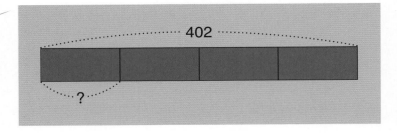

In this example, we know there are 4 in each group. We are looking for the number of groups.

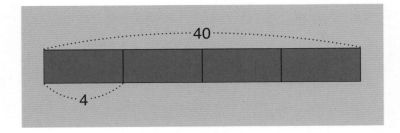

Teacher's Guide 2A Chapter 7
© 2017 Singapore Math Inc.

At this stage, students do not need to draw bar models, and a strategy for using sticky notes is included in the lesson. Students could also use linking cubes or drawings of simple shapes to represent equal groups.

In grade 3, students will be required to draw the models as they solve more complex problems.

Materials

- Counters
- Index cards
- Construction paper (optional)
- Multiplication fact cards made by students in this chapter
- Paper plates
- Sticky notes
- Whiteboards

Blackline Masters

- Array Dot Card — 2s
- Array Dot Card — 5s
- Array Dot Card — 10s
- Division by 2 Fact Cards
- Kaboom Cards
- Multiplication Chart — 2
- Multiplication Chart — 5
- Multiplication Chart — 10
- Multiplication Squares 2–5–10 Game Board

Storybooks

- *Lots of Ladybugs* by Michael Dahl
- *Count on Pablo* by Barbara deRubertis
- *Minnie's Diner* by Dayle Ann Dodds
- *The Great Divide* by Dayle Ann Dodds
- *How Many Feet in the Bed?* by Diane Johnston Hamm
- *2 × 2 = Boo!: A Set of Spooky Multiplication Stories* by Loreen Leedy
- *Arctic Fives Arrive* by Elinor J. Pinczes

Activities

Games and activities included in this chapter are designed to provide practice with multiplication and division of 5, 2, and 10. They can be used after students complete the **Do** questions, or anytime additional practice is needed. Students should know their multiplication facts for 5, 2, and 10 from memory by the end of the chapter.

Chapter Opener

Objective

- Investigate multiplication and division.

Materials

- Counters
- Books from the list on page 190 of this Teacher's Guide

Provide pairs of students with counters and have them find the number of each type of fruit. Ask students what they notice about the numbers of fruit.

Have them write and share some multiplication or division equations from objects they see in the textbook, classroom, or from one of the books listed on page 190 of this Teacher's Guide.

Take the opportunity to reinforce skip counting by 5, 2, and 10 with one of the activities.

Activities

▲ Hundred Chart Patterns

Materials: Hundred charts in dry erase sleeves, dry erase markers

Have students color in the numbers said when skip counting by 5. Have them look for patterns in the colored numbers.

Repeat for skip counting by 2 and 10.

▲ Catch and Count

Materials: Soft ball or bean bag

Students toss the ball to each other. The player who catches the ball says the next number when counting by 5.

The first player starts with, "Zero," then tosses the ball. The catcher says, "Five," then tosses the ball again, and so on. Count back from 100 by 5 for more of a challenge.

Repeat for skip counting by 2 and 10.

Lesson 1 The Multiplication Table of 5

Objectives

- Build and understand the structure of the multiplication table of 5.
- Look for patterns in the multiplication table of 5.

Lesson Materials

- Counters, 50 per student
- Multiplication Chart — 5 (BLM)
- Array Dot Card — 5s (BLM)

Think

Pose the **Think** problem with the mangoes and tell students we are going to find the total amount of mangoes in more than one box.

Provide students with counters and a Multiplication Chart — 5 (BLM).

Have students start by making one row of 5 counters and filling in the chart:

- 1 box, with 5 mangoes in each box, makes a total of 5 mangoes. The equation column can be left blank.

Reinforce the term "array" to describe the way the counters are laid out.

Have students add a second row of 5 to see how many mangoes there are in 2 boxes.

- 2 boxes, with 5 mangoes in each box, makes a total of 10 mangoes. $2 \times 5 = 10$

Continue to add rows and complete the chart.

Ask students what they notice about the total numbers of mangoes and discuss Emma's comment on the increasing total. Students should notice that as the number of groups increases by 1, the total number of mangoes increases by 5.

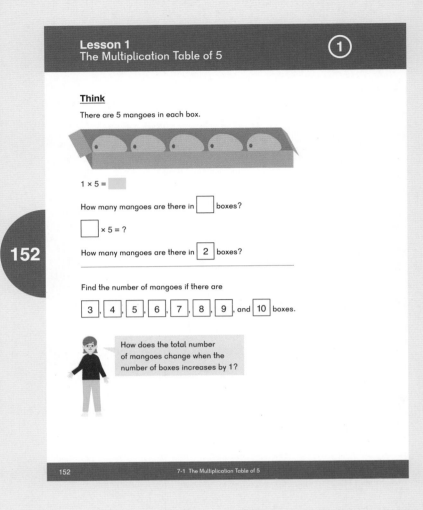

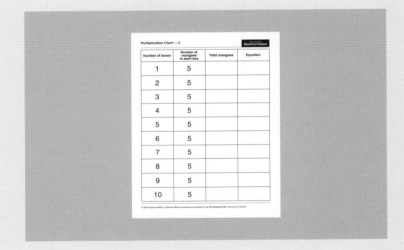

Teacher's Guide 2A Chapter 7 © 2017 Singapore Math Inc.

Learn

Have students add the equations to their charts.

Multiplication Chart — 5			Dimensions Math Blackline Masters
Number of boxes	Number of mangoes in each box	Total mangoes	Equation
1	5	5	1 × 5 = 5
2	5	10	2 × 5 = 10
3	5	15	3 × 5 = 15
4	5	20	4 × 5 = 20

Discuss the textbook examples with the blue boxes. The equations help students understand they can figure out unknown facts from facts they already know.

Emma introduces the term "product" as the term for the total in a multiplication equation. As the number of groups increases by 1, the product increases by 5.

Have students look for the pattern in the ones digits of the product and discuss Alex's thought. Multiples of five always have a 0 or a 5 in the ones place.

Do

❶ In (a) and (b), students should notice that 4 is double 2, so 4 × 5 is double 2 × 5.

❷ Use the activity below and have students create their own dot paper. Alternatively, use Array Dot Cards — 5s (BLM).

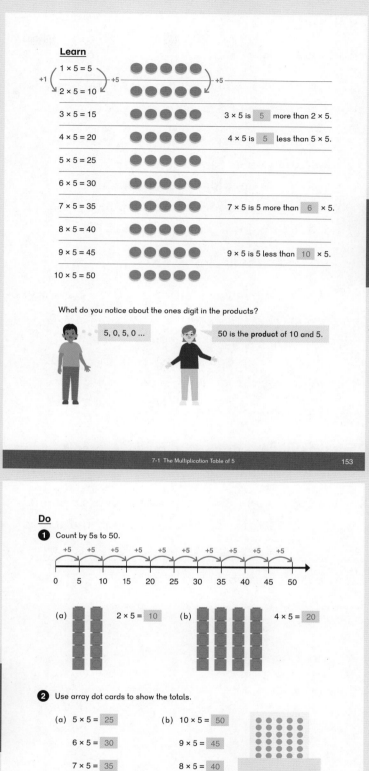

Learn

1 × 5 = 5
2 × 5 = 10
3 × 5 = 15   3 × 5 is 5 more than 2 × 5.
4 × 5 = 20   4 × 5 is 5 less than 5 × 5.
5 × 5 = 25
6 × 5 = 30
7 × 5 = 35   7 × 5 is 5 more than 6 × 5.
8 × 5 = 40
9 × 5 = 45   9 × 5 is 5 less than 10 × 5.
10 × 5 = 50

What do you notice about the ones digit in the products?

5, 0, 5, 0 ...

50 is the **product** of 10 and 5.

7-1 The Multiplication Table of 5   153

Do

❶ Count by 5s to 50.

0 5 10 15 20 25 30 35 40 45 50

(a)   2 × 5 = 10   (b)   4 × 5 = 20

❷ Use array dot cards to show the totals.

(a) 5 × 5 = 25   (b) 10 × 5 = 50
   6 × 5 = 30   9 × 5 = 45
   7 × 5 = 35   8 × 5 = 40

❸ A nickel is worth 5 cents.
How many cents are 3 nickels worth?

 3 × 5 = 15

3 nickels are worth 15 ¢.

Exercise 1 · page 149

154   7-1 The Multiplication Table of 5

Activities

▲ Choral Counting

Using your thumb pointing up or down, have students chorally count up and down by fives. Example:

- "Let's count by fives starting at 0. First number?" Class responds, "0."
- Point thumb up, class responds, "5."
- Point thumb up again, class responds, "10."
- Point thumb down, class responds, "5."

Start at random multiples of 5:

- "Let's count by fives starting at 20. First number?" Class, "20."
- Point thumb up, class responds, "25."
- Point thumb up again, class responds, "30."
- Point thumb down, class responds, "25."

▲ Array Dot Cards

Materials: Index cards, craft punches/hole punches or dot stickers, glue

Have each student use either dot stickers or punches to make a 5 × 10 array on index cards for future reference.

Students will be making similar cards for × 2 and × 10 later in this chapter.

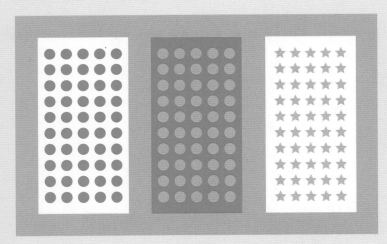

Exercise 1 · page 149

Lesson 2 Multiplication Facts of 5

Objectives

- Understand the commutative property in the multiplication table of 5.
- Learn the multiplication facts of 5.

Materials

- Counters
- Index cards, 20 per student, or construction paper

Think

Have students model the **Think** problem with counters and share their solutions.

Learn

Sofia reminds us that the order students multiply will not change the total. Whether it is 4 groups of 5 or 5 groups of 4, the total is the same.

While learning the facts, the commutative property will become very valuable. When tables of 6 to 9 are taught in **Dimensions Math® 3A**, students who know 5 × 6 will also know 6 × 5. Thus, students will know some facts for groups of 6 before they count by 6.

This means there are fewer facts that need to be learned in each successive table.

This part of the lesson may progress quickly. Students will need time for the **Do** part of the lesson to create flash cards.

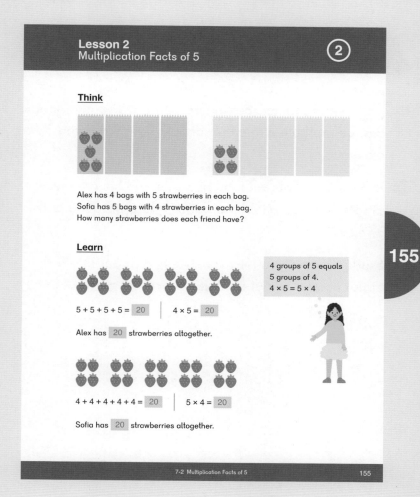

Do

❶—❷ Reiterate that it does not matter which number represents the number of groups and which represents the number in each group. The numbers can be multiplied in either order to get the same answer.

❺ Provide students with index cards and have them create their own flash cards for future practice and games.

Students can also fold construction paper into 8 equal parts and cut out their own flash cards.

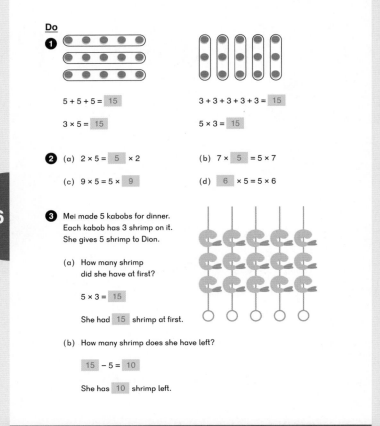

156

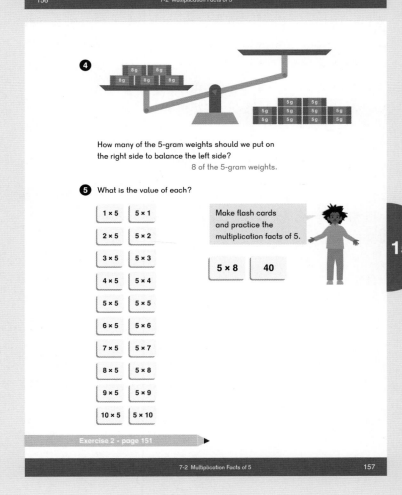

157

Activities

▲ Multiplication Wheels

Materials: Paper plates with the center cut out

Create several multiplication wheels with the numbers 1 to 10 in random order as "spokes" along the edge of the paper plates.

Students lay the wheel on a whiteboard and write the number they are multiplying by in the center of the wheel. (In this lesson it is 5.)

Students multiply the number on the spoke and the number in the center, and write the product on the whiteboard, outside of the wheel.

▲ Five's a Hopping!

Materials: Sidewalk chalk or painter's tape, multiplication by 5 cards made in this lesson, fly swatters or bean bags

Create two grids like the one shown below, using either chalk outside or painter's tape inside.

One student is the Caller. Two students are the Hoppers and stand on their home squares. The Caller flips over a multiplication by 5 card and calls out the equation.

Hoppers must hop on the answer.

The first Hopper who misses the correct square becomes the next Caller. (Include a non-multiple of 5 in the extra square.)

Five's a Bopping This mini indoor game could be played with a smaller grid on paper and fly swatters to smack numbers, or with bean bags to toss onto numbers.

5	20	35
30	10	15
40	25	45
50	17	Home

◀ **Exercise 2 · page 151**

Lesson 3 Practice A

Objective

- Practice the multiplication facts of 5.

Materials

- Multiplication Facts for 5 flash cards created in Lesson 2

Practice

After students complete the **Practice** in the textbook, have them continue to practice multiplication facts for 5 with activities and flash cards from the chapter.

Students should continue to practice these facts until they know them from memory.

5 — **7** These questions can be more challenging as they apply multiplication by 5 through measurement. In **7**, Dexter has 7 nickels, and the value of the 7 nickels is 35 cents.

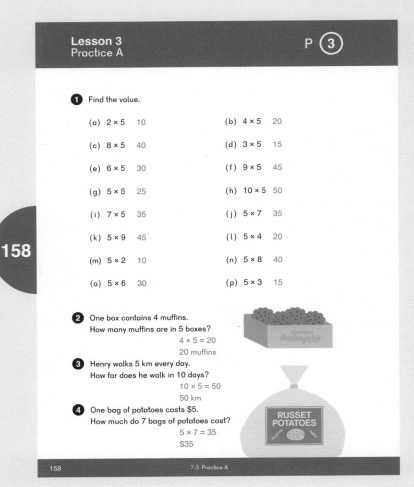

Lesson 3 Practice A   P **3**

1 Find the value.

(a) 2 × 5 10	(b) 4 × 5 20
(c) 8 × 5 40	(d) 3 × 5 15
(e) 6 × 5 30	(f) 9 × 5 45
(g) 5 × 5 25	(h) 10 × 5 50
(i) 7 × 5 35	(j) 5 × 7 35
(k) 5 × 9 45	(l) 5 × 4 20
(m) 5 × 2 10	(n) 5 × 8 40
(o) 5 × 6 30	(p) 5 × 3 15

158

2 One box contains 4 muffins. How many muffins are in 5 boxes?
4 × 5 = 20
20 muffins

3 Henry walks 5 km every day. How far does he walk in 10 days?
10 × 5 = 50
50 km

4 One bag of potatoes costs $5. How much do 7 bags of potatoes cost?
5 × 7 = 35
$35

158   7-3 Practice A

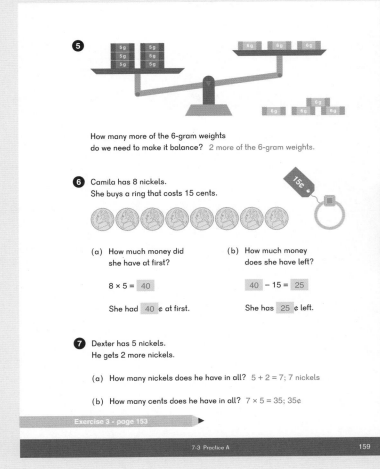

5

How many more of the 6-gram weights do we need to make it balance?   2 more of the 6-gram weights.

6 Camila has 8 nickels. She buys a ring that costs 15 cents.

(a) How much money did she have at first?

8 × 5 = 40

She had 40 ¢ at first.

(b) How much money does she have left?

40 − 15 = 25

She has 25 ¢ left.

159

7 Dexter has 5 nickels. He gets 2 more nickels.

(a) How many nickels does he have in all?   5 + 2 = 7; 7 nickels

(b) How many cents does he have in all?   7 × 5 = 35; 35¢

Exercise 3 - page 153

7-3 Practice A   159

Activities

▲ Hopscotch × 5

Materials: Pebble or other marker for each player, chalk or paper plates and painter's tape

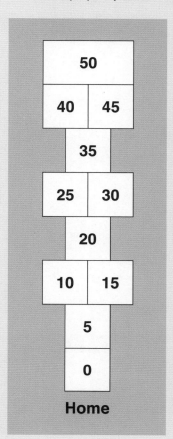

Home

Play outside or in a gym. Draw a hopscotch board using chalk, or tape down paper plates to create a hopscotch board with 0 as the starting spot and the numbers in the squares increasing by 5.

Players take turns standing in the 0 square (Home) and tossing the marker. On their first turn, players aim for the 5 square. On each turn, players hop over the square with the marker and continue hopping in order, saying the numbers in each square aloud.

Square 50 is a rest stop. Players can put both feet down before turning around and hopping back to 0.

Players pause in square 10 to pick up the marker, hop in square 5, and out. On his next turn, the player aims his marker for square 10, etc.

A player's turn is over if:

- His marker does not land in the correct square.
- He loses his balance and puts a second foot or a hand down.
- He lands in a square where the marker is.

The winner is the first player to get through all 10 turns.

▲ Nim

Materials: Counters, geared clock

Nim is a logic game that is as easy to play as tic-tac-toe, but requires more complex thinking. Versions of **Classic Nim** and **Poison** were introduced in **Chapter 4**.

Clock Nim: In this version, use a geared clock in place of counters. Beginning with the hands on the clock at 12:00, players take turns moving the minute hand clockwise 5, 10, or 15 minutes. The player who moves the hands to 12:50 exactly is the winner.

 Exercise 3 • page 153

Lesson 4 The Multiplication Table of 2

Objectives

- Build and understand the structure of the multiplication table of 2.
- Look for patterns in the multiplication table of 2.
- Multiply numbers within the multiplication table of 2.

Materials

- Counters, 20 per student
- Multiplication Chart — 2 (BLM)
- Array Dot Card — 2s (BLM)

Think

Pose the **Think** problem with the watermelons and tell students they are going to find the total amounts in more than one box.

Provide students with counters and Multiplication Chart — 2 (BLM).

Have students start by making one row of 2 counters and filling in the chart:

- 1 box, with 2 watermelons in each box, makes a total of 2 watermelons.

Have students add a second row of 2 to see how many watermelons there are in 2 boxes.

Continue to add rows and complete the chart.

Discuss Mei's question on the increasing total.

Students should notice that as the number of groups increases by 1, the product increases by 2.

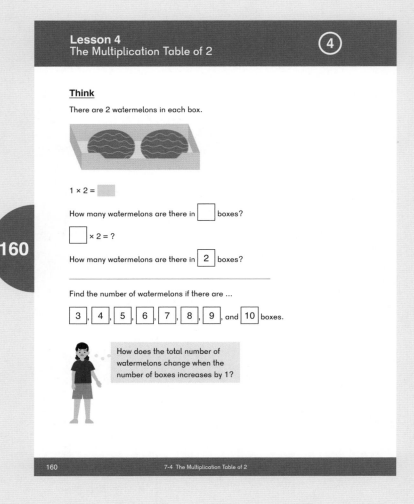

Ask students, "How is this problem different from the problem with the mangoes where we counted by fives?"

- We are multiplying 2, not 5.
- The total is increasing by 2.

Then ask, "How is it the same?"

- We are still making equal groups.

Learn

Have students add the equations to their charts.

Discuss the examples in the textbook with the blue boxes. The equations help students understand that they can figure out unknown facts from facts that they already know.

Discuss Emma's comments. Students have not been introduced to even numbers, but should notice a pattern of 2, 4, 6, 8, 0, 2, 4, 6, 8, 0. Even and odd numbers are formally introduced in **Dimensions Math® 3A**.

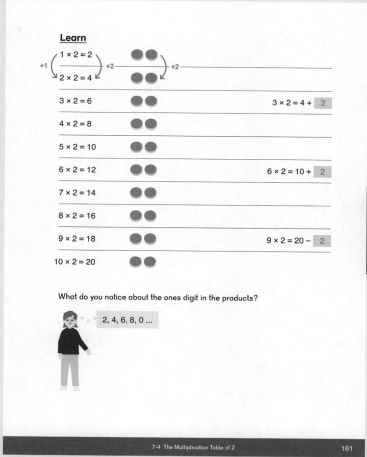

Do

3 Repeat the activity **Array Dot Cards** from Lesson 1 for the multiplication facts for 2 to create a 2s table, or use the Array Dot Cards — 2 (BLM).

Activity

▲ Hopscotch × 2

Materials: Pebble or other marker for each player, sidewalk chalk or paper plates and painter's tape

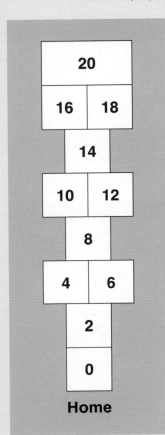

Home

For Hopscotch × 2, create a hopscotch with multiples of 2.

The first player stands in the 0 (home) square and tosses her marker to square 2.

The player hops over square 2 to square 4 and then continues hopping in order, skipping any squares with markers and saying the numbers in each square aloud.

Exercise 4 • page 157

162

Do

1 Count by 2s to 20.

+2 +2 +2 +2 +2 +2 +2 +2 +2 +2

0　2　4　6　8　10　12　14　16　18　20

2 (a) 2 × 2 = 4　　(b) 4 × 2 = 8

3 Use array dot cards to show the totals.

(a) 5 × 2 = 10　　(b) 10 × 2 = 20

6 × 2 = 12　　9 × 2 = 18

7 × 2 = 14　　8 × 2 = 16

4 8 children each have $2.
How much money do they have altogether?

8 × 2 = 16　　|　They have $ 16 .

5 2 people are in each kayak.
How many people are in 7 kayaks?

7 × 2 = 14
14 people

Exercise 4 • page 157

162　　　　7-4 The Multiplication Table of 2

Lesson 5 Multiplication Facts of 2

Objectives

- Understand the commutative property in the multiplication table of 2.
- Learn the multiplication facts of 2.

Materials

- Counters
- Index cards, 20 per student, or construction paper

Think

Pose the **Think** problem with the chocolates. At this point, most students will know that Emma and Mei each have 12 chocolates in all. Give students counters to represent the chocolates and have them compare the two sets of boxes.

Learn

Alex reminds students that finding 6 × 2 by adding 6 + 6, instead of 2 + 2 + 2 + 2 + 2 + 2, is simpler even if the muffins are in groups of 2.

Ask, "What fact in the multiplication table of 2 did we already learn when we learned the multiplication table of 5?" (5 × 2 = 2 × 5)

As with multiplication by 5, this part of the lesson may progress quickly. Students will need time in the **Do** part of the lesson to create more flash cards.

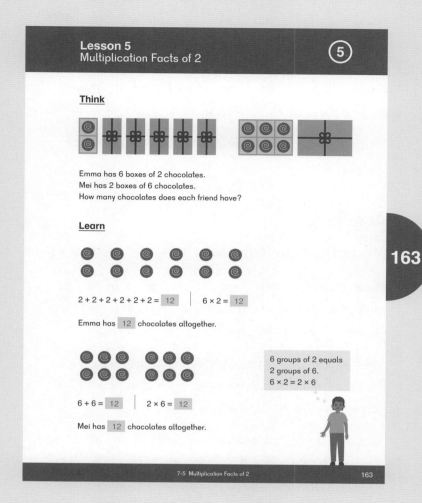

Lesson 5
Multiplication Facts of 2

⑤

Think

Emma has 6 boxes of 2 chocolates.
Mei has 2 boxes of 6 chocolates.
How many chocolates does each friend have?

Learn

2 + 2 + 2 + 2 + 2 + 2 = 12   6 × 2 = 12

Emma has 12 chocolates altogether.

6 groups of 2 equals 2 groups of 6.
6 × 2 = 2 × 6

6 + 6 = 12   2 × 6 = 12

Mei has 12 chocolates altogether.

7-5 Multiplication Facts of 2   163

163

Do

⑤ Provide students with index cards and have them create their own flash cards for future practice and games.

Students can also fold construction paper into 8 equal parts and cut out their own flash cards.

Activities

▲ Multiplication Kaboom

Materials: Kaboom Cards (BLM), several sets of Multiplication Fact Cards for 2 and 5 created by students in this chapter

Shuffle and place the cards in a pile, facedown. Players take turns drawing a card and saying the answer to the multiplication fact.

Students keep the cards they answer correctly, and return the ones that they answer incorrectly. When students draw a Kaboom Card (BLM), they must return all of their collected cards to the pile.

The player with the most cards at the end of the time limit is the winner.

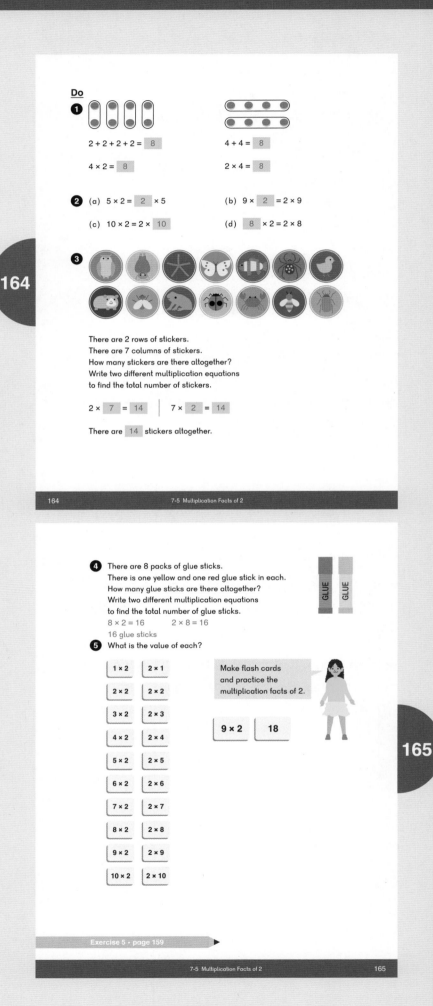

▲ Two's a Hopping!

Materials: Sidewalk chalk or painter's tape, Multiplication Fact Cards for 2 created in lesson

Create two grids like the one shown below, using either chalk outside or painter's tape inside.

One student is the Caller. Two students are the Hoppers and stand on their home squares. The Caller flips over a multiplication by 2 card and calls out the equation.

Hoppers must hop on the answer.

The first Hopper who misses the correct square becomes the next Caller. (Include a non-multiple of 2 in the extra square.)

Two's a Bopping

This mini indoor game could be played with a smaller grid on paper and fly swatters to smack numbers, or with bean bags to toss onto numbers.

8	20	4
6	2	18
14	25	12
16	10	Home

Exercise 5 • page 159

Lesson 6 Practice B

Objective

- Practice multiplication of 2 and 5.

Materials

- Multiplication Facts Flash cards created in this chapter

Practice

After students complete the **Practice** in the textbook, have them continue to practice multiplication facts for 2 and 5 with activities and flash cards from the chapter.

Students should continue to practice these facts until they know them from memory.

2—**3** Have students discuss how they found their answers.

6—**7** Encourage students to draw pictures to help solve the problems.

7 Although a picture is given, students may benefit from drawing it themselves.

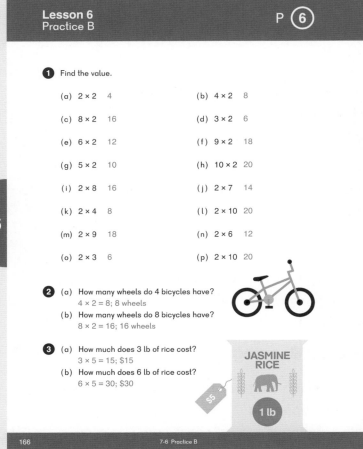

Lesson 6
Practice B
P 6

1 Find the value.

(a) 2 × 2 4	(b) 4 × 2 8
(c) 8 × 2 16	(d) 3 × 2 6
(e) 6 × 2 12	(f) 9 × 2 18
(g) 5 × 2 10	(h) 10 × 2 20
(i) 2 × 8 16	(j) 2 × 7 14
(k) 2 × 4 8	(l) 2 × 10 20
(m) 2 × 9 18	(n) 2 × 6 12
(o) 2 × 3 6	(p) 2 × 10 20

166

2 (a) How many wheels do 4 bicycles have?
   4 × 2 = 8; 8 wheels
   (b) How many wheels do 8 bicycles have?
   8 × 2 = 16; 16 wheels

3 (a) How much does 3 lb of rice cost?
   3 × 5 = 15; $15
   (b) How much does 6 lb of rice cost?
   6 × 5 = 30; $30

JASMINE RICE
$5
1 lb

166   7-6 Practice B

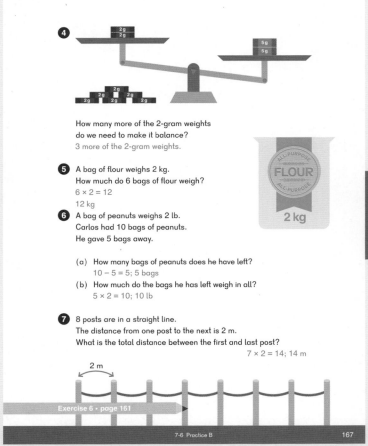

4 How many more of the 2-gram weights do we need to make it balance?
3 more of the 2-gram weights.

5 A bag of flour weighs 2 kg.
How much do 6 bags of flour weigh?
6 × 2 = 12
12 kg

6 A bag of peanuts weighs 2 lb.
Carlos had 10 bags of peanuts.
He gave 5 bags away.

(a) How many bags of peanuts does he have left?
10 − 5 = 5; 5 bags
(b) How much do the bags he has left weigh in all?
5 × 2 = 10; 10 lb

7 8 posts are in a straight line.
The distance from one post to the next is 2 m.
What is the total distance between the first and last post?
7 × 2 = 14; 14 m

FLOUR
2 kg

167

2 m

Exercise 6 · page 161

7-6 Practice B   167

Activity

★ **Multiply Team Race**

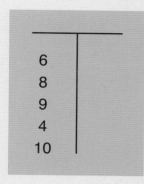

	× 5
6	30
8	40
9	45
4	20
10	50

Students seeking a challenge will enjoy the race element of this game. Have 2–4 students go to the board and make a large "T" shape. Randomly call out 5 numbers from 1 to 10 and have students write them on the left side of the T.

Then give them a number to multiply by, for example, "times 5" in this scenario.

Students write "× 5" on the top of the T, then proceed to solve the problems on the right side as fast as they can.

Exercise 6 • page 161 ▶

Lesson 7 The Multiplication Table of 10

Objectives

- Build and understand the structure of the multiplication table of 10.
- Look for patterns in the multiplication table of 10.
- Learn the multiplication facts of 10.

Materials

- Counters, 100 per student
- Multiplication Chart — 10 (BLM)
- Array Dot Card — 10s (BLM)
- Index cards, 20 per student or paper to create flash cards

Think

Pose the **Think** problem with the walnuts and tell students we are going to find the total amounts in more than one box.

Provide each student with 100 counters and Multiplication Chart — 10 (BLM).

Have students start by making one row of 10 counters and filling in the chart:

- 1 bag with 10 walnuts in each bag makes a total of 10 walnuts.

Have students add a second row of 10 to see how many walnuts there are in 2 bags.

Continue to add rows and complete the chart.

Discuss Dion's thought on the increasing total.

Students should notice that as the number of groups increases by 1, the product increases by 10.

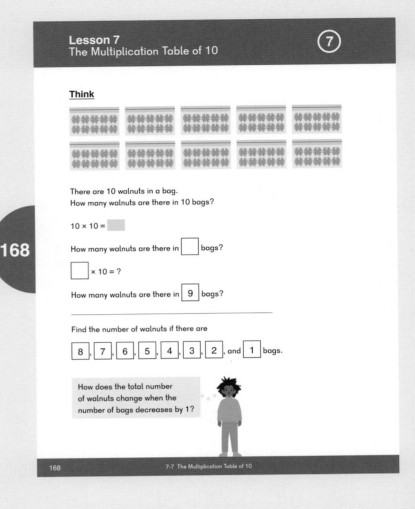

168

Ask students, "How is this problem different from the problem with the mangoes where we counted by 5, or the watermelons where we counted by 2?"

- We are multiplying 10.
- The numbers are increasing by 10.

Then ask, "How is it the same?"

- We are still making equal groups.

Learn

Have students add the equations to their charts.

Discuss Mei's comments. Multiples of 10 always have a 0 in the ones place.

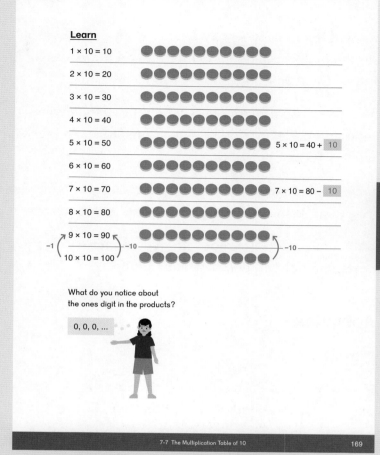

Do

❸ Repeat the activity **Array Dot Cards** from Lesson 1 for the multiplication facts for 10 to create a 10s table, or use the Array Dot Cards — 10 (BLM).

❺ This is a good problem to point out the difficulty in repeated addition. Adding 4 ten times takes longer than adding 10 four times.

❻ Provide students with index cards and have them create their own flash cards for future practice and games.

Students can also fold construction paper into 8 equal parts and cut out their own flash cards.

Activities

▲ Match

Materials: Student-made array dot cards for 2, 5, and 10, student-made flash cards for 2, 5, and 10

Students can use their flash cards and fact cards to practice their facts.

Students mix one set of each of the cards and lay out 20 cards, faceup. The rest of the cards are kept as a draw pile.

When they have a match, they say the fact and collect the two cards that match. When they collect their two cards that match, the empty spots are filled with two more cards from the draw pile.

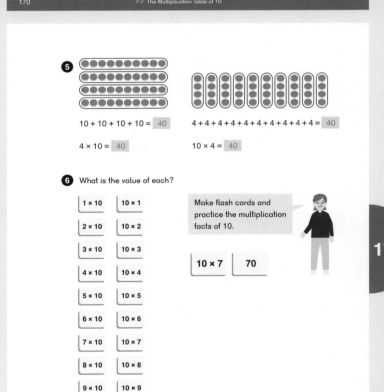

▲ Multiplication Squares

Materials: Multiplication Squares 2–5–10 Game Board (BLM), deck of Number Cards (BLM) 1 to 10 or playing cards 1 to 10, modified die with sides labeled 2, 2, 5, 5, 10, 10

Use game board in dry erase sleeves with colored markers or paper with colored pencils.

On each turn, a player rolls the die, draws a card, and multiplies the numbers together to find the product.

The player connects the dots to make a line on one side of her product.

The goal is to "capture" the product by boxing it in. The player who completes the square shades it in and rolls again.

Play continues until all boxes have been shaded in. The player with the most "captured" squares is the winner.

If there are no places to play on the board, the player loses a turn.

Exercise 7 • page 165 ▶

Objective

- Use a related multiplication sentence to solve division problems without remainder where the divisor is 2.

Materials

- Counters
- Paper plates

Think

Provide students with counters and paper plates and adequate time to work through the **Think** problem with the strawberries.

Have students share their strategies.

Learn

Have students discuss textbook page 172. Ask them if this is a sharing or grouping problem. Sofia points out that, in division, we can find how many items are in each group if we know the total and the number of groups.

When we multiply, we know the number of groups and how many are in each group. We find the total or product.

In division, we either:

- Know the total and the number of groups, and need to find how many in each group (sharing) or,
- Know the total and the number in each group, and need to find the number of groups (grouping).

The **Think** problem is a sharing situation.

Discuss Dion's comments in the textbook. Point out that if students know their multiplication facts, they can use them to find division facts.

Note: Students will use the term "quotient" in **Dimensions Math® 3A**.

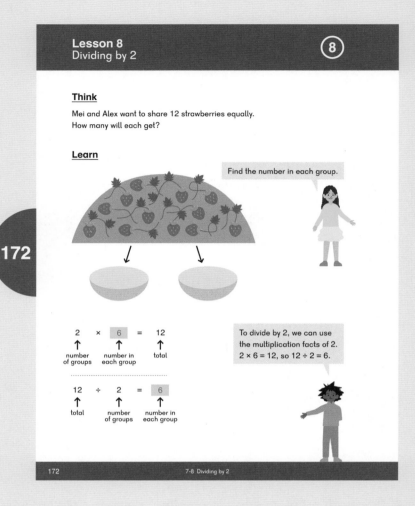

Do

① Have students discuss the different situations. Have them show the two situations with a drawing.

In (a) there are 2 groups, in (b) there are 8 groups.

③ Have students share a related multiplication equation.

⑥ Have students draw simple pictures to solve this problem as needed.

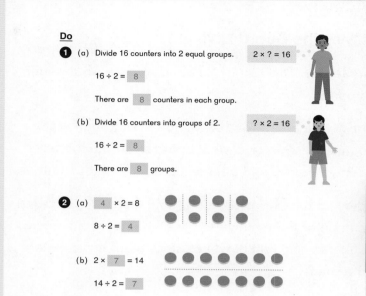

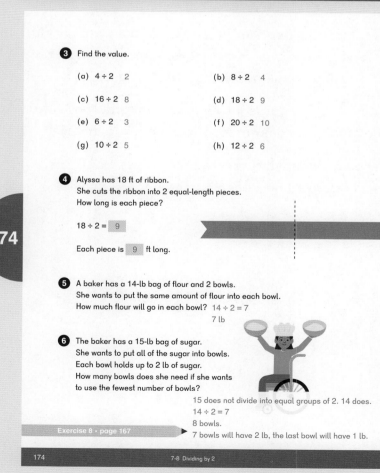

Activities

▲ Clear the Board Division

Materials: Division by 2 Fact Cards (BLM) for each student, 7 counters

Have students write the numbers 1 through 10 on a number path on a whiteboard to create their own game boards, and put counters on 7 of the numbers.

Players take turns drawing fact cards and removing a counter if the answer on their number path has a counter. The first player to clear all of her counters is the winner.

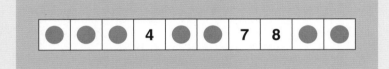

▲ Division Kaboom

Materials: Kaboom Cards (BLM), several sets of Division by 2 Fact Cards (BLM)

Shuffle and place the Division by 2 Fact Cards (BLM) in a pile, facedown. Players take turns drawing a card and saying the answer to the division fact.

They keep the cards they answer correctly, and return the cards they answer incorrectly back to the pile.

When a student draws a Kaboom Card (BLM), he must return all his collected cards to the pile.

The player with the most cards at the end of the time limit is the winner.

Exercise 8 · page 167

Objective

- Use a related multiplication sentence to solve division problems without remainder where the divisor is 5 or 10.

Lesson Materials

- Counters
- Paper plates

Think

Provide students with counters and paper plates to work through the **Think** problem.

Learn

Ask students, "How is this problem different from the previous lesson with the strawberries?"

- We are dividing by 5, not by 2
- We know the total and the number in each group, and need to find how many groups there are.

Then ask, "How is it the same?"

- We can still use multiplication facts to find an answer to a division equation.

Ask if this is a sharing problem or a grouping problem. (Grouping)

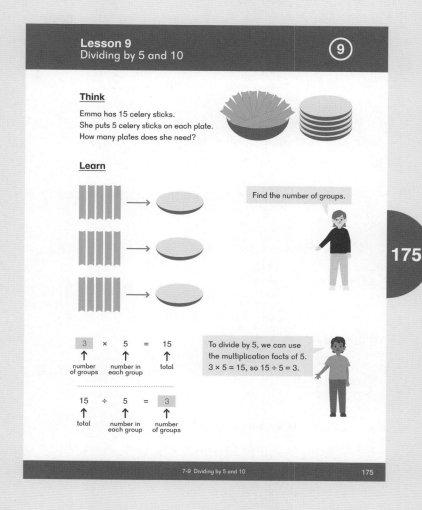

Lesson 9
Dividing by 5 and 10 — ⑨

Think

Emma has 15 celery sticks.
She puts 5 celery sticks on each plate.
How many plates does she need?

Learn

Find the number of groups.

175

3	×	5	=	15

number of groups — number in each group — total

To divide by 5, we can use the multiplication facts of 5.
3 × 5 = 15, so 15 ÷ 5 = 3.

15	÷	5	=	3

total — number in each group — number of groups

7-9 Dividing by 5 and 10 — 175

Do

1 Students should use counters and discuss their solutions. The friends are reminding us of a related multiplication fact.

Have students use counters if needed. Most should be working with just the pages in the textbook.

3 Have students share a related multiplication equation.

Activity

▲ Leftovers

Materials: 45 counters, modified die with sides labeled 2, 2, 5, 5, 10, 10

Player One rolls the die and divides her counters by the number on the die.

For example, Player One rolls a 2. She divides the counters into 2 equal groups with 1 left over. That player keeps the leftover counter and play continues with the remaining 44 counters.

Player Two rolls a 10 and divides the remaining 44 counters by 10. She has 10 groups of 4, with 4 counters left over. Player Two keeps the 4 counters and returns the 40 remaining counters.

When no more divisions can be made, the game is over. The player with the most counters is the winner.

Exercise 9 • page 171

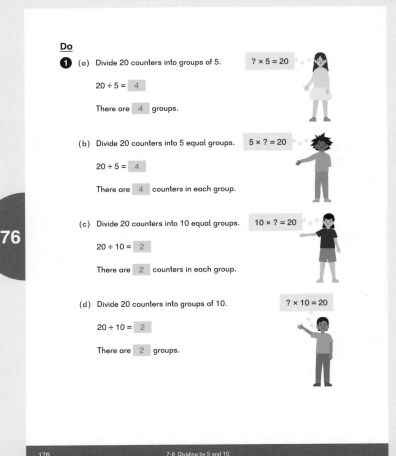

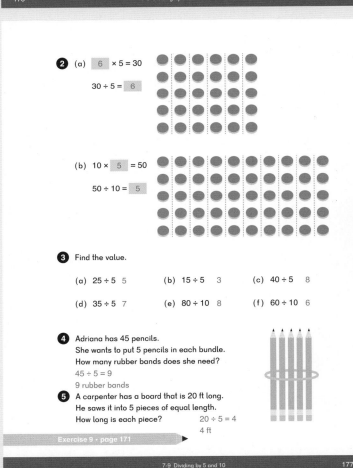

Objective

- Practice multiplication and division of 2, 5, and 10.

Practice

After students complete the **Practice** in the textbook, have them continue to practice multiplication facts for 2, 5, and 10. Students should continue to practice these facts until they know them from memory.

5 Ask students if they have learned a fact that can help them solve the problem. You may need to point out that there can be some beads left over.

Activity

▲ **Five and Ten Hopping!**

Materials: Sidewalk chalk or painter's tape, fact cards for 5 and 10

Create two grids like the one shown to the right, using either chalk outside or painter's tape inside.

One student is the Caller. Two students are the Hoppers and stand on their home squares. The Caller flips over a multiplication by 5 card and calls out the equation.

Hoppers must hop on the answer to the fact. Note that the ten square will work for 2 x 5 and 1 x 10.

The first Hopper who misses the correct square becomes the next Caller.

Five and Ten a Bopping can also be played as an indoor mini game. See instructions in Lesson 5 on page 205 of this Teacher's Guide.

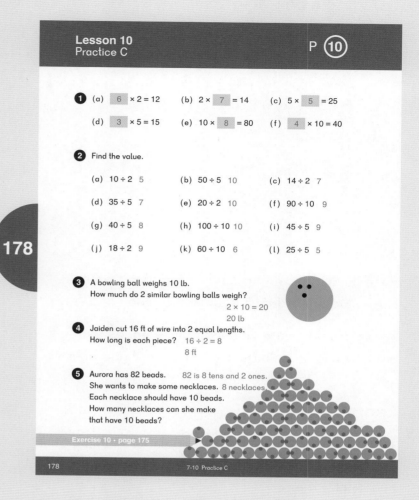

5	20	60	35
30	10	15	70
40	25	45	80
50	90	100	Home

Exercise 10 • page 175

Lesson 11 Word Problems

Objectives

- Solve problems involving multiplication and division.
- Determine the correct operation to solve a multiplication or division problem.

Materials

- Sticky notes

Think

Pose the **Think** word problems from the textbook. Have students write an equation for each and solve.

These problems are limited to one step, as the emphasis is on determining the correct operation.

Learn

To introduce students to bar models for multiplication and division, provide sticky notes and have them use their whiteboards.

Sticky notes help demonstrate that each bar is an equal amount (or unit) and represent the same amount in a problem. The whiteboard allows multiple problems to be solved by rearranging the bars (sticky notes), numbers, and question mark.

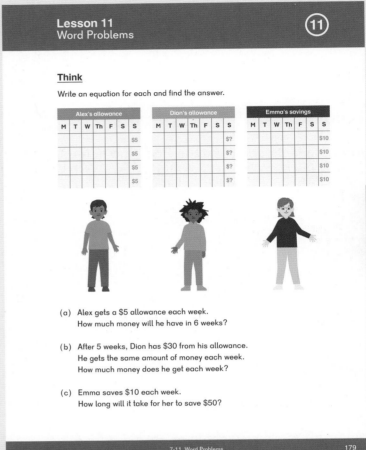

Lesson 11
Word Problems

(11)

Think

Write an equation for each and find the answer.

Alex's allowance						
M	T	W	Th	F	S	S
						$5
						$5
						$5
						$5

Dion's allowance						
M	T	W	Th	F	S	S
						$?
						$?
						$?
						$?

Emma's savings						
M	T	W	Th	F	S	S
						$10
						$10
						$10
						$10

(a) Alex gets a $5 allowance each week. How much money will he have in 6 weeks?

(b) After 5 weeks, Dion has $30 from his allowance. He gets the same amount of money each week. How much money does he get each week?

(c) Emma saves $10 each week. How long will it take for her to save $50?

Teacher's Guide 2A Chapter 7 © 2017 Singapore Math Inc.

In (a), students create a model that shows 6 weeks in which Alex gets $5. To show how much money Alex has after 6 weeks, they can add a "?" as shown below.

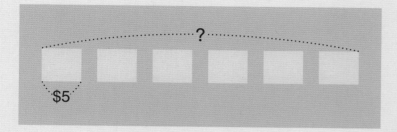

Dion's model starts with the whole and divides that into 5 equal parts:

Ask students if (b) and (c) are sharing or grouping situations.

(c) Challenge students to think about how this model differs from the model in (b). Students know the whole of $50 is what Emma needs to save. We know Emma saves $10 each week, but don't know how many weeks it will take her to save $50.

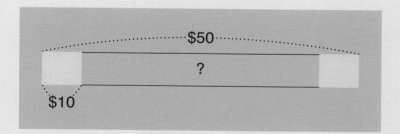

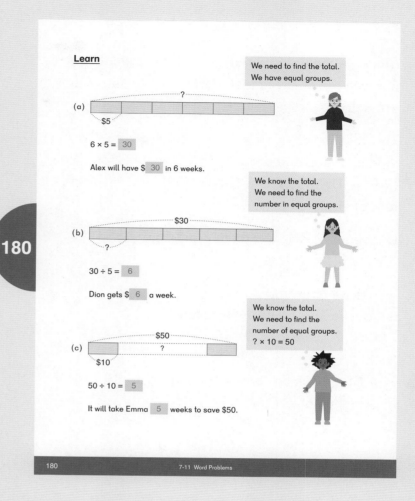

Learn

(a)

We need to find the total. We have equal groups.

$6 \times 5 = $ 30

Alex will have $ 30 in 6 weeks.

We know the total. We need to find the number in equal groups.

(b)

$30 \div 5 = $ 6

Dion gets $ 6 a week.

We know the total. We need to find the number of equal groups. ? × 10 = 50

(c)

$50 \div 10 = $ 5

It will take Emma 5 weeks to save $50.

Ask students if this is a grouping or sharing situation. (Grouping)

For a division by grouping bar model, students won't know how many units are needed. While Emma's problem is simple, students will be solving problems like this one in the future:

Cole has $462. Each week he spends $4 on coffee. How many weeks can Cole buy coffee before he runs out of money?

Do

Students are not required to draw the models but can use the sticky notes or draw the models if they are helpful.

This lesson is limited to problems involving multiplication and division. In the next lesson (Review B), students will solve problems involving all four operations.

Questions to ask about the problems:

- Is a total given?
- Do we know the parts?
- Do we know a part and a total?
- What do we know?

❸ — ❹ Students can draw a bar model or a simple drawing to solve.

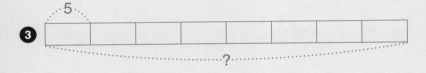

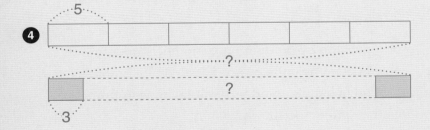

Exercise 11 · page 177

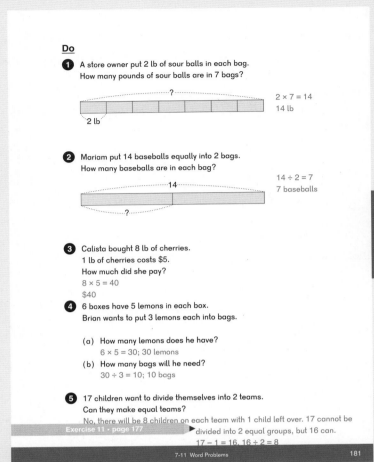

Review 2

Objective

- Review topics from Chapter 1 through Chapter 7.

Use this cumulative review as necessary to practice and reinforce content and skills from the first seven chapters.

Activity

▲ Snowball Review

Materials: Regular paper

Write multiplication and division problems on sheets (or half sheets) of paper. Make enough to give each student 2 or 3 problems.

Have each student crumple up the paper and have a classroom snowball fight for one minute. (No running, safety first!)

At the end of the fight, ask kids to grab a snowball and return to their seats. One at a time, have students unwrap the paper, show the class, and state the answer to the problem.

When each student has solved one fact problem, have another fight and repeat with remaining snowballs on the floor.

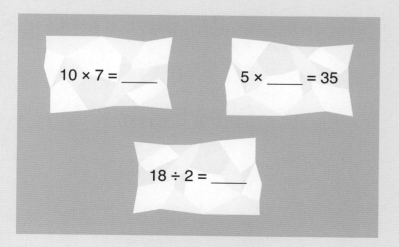

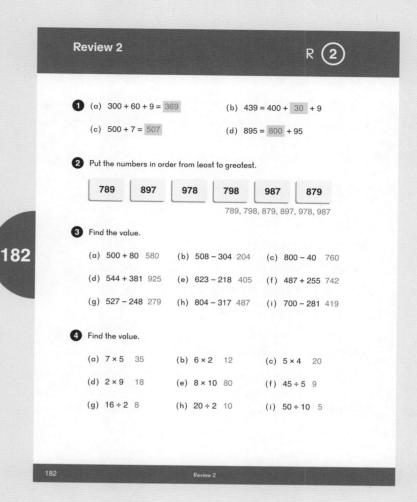

Review 2 R ②

1 (a) $300 + 60 + 9 =$ 369   (b) $439 = 400 +$ 30 $+ 9$

(c) $500 + 7 =$ 507   (d) $895 =$ 800 $+ 95$

2 Put the numbers in order from least to greatest.

| 789 | 897 | 978 | 798 | 987 | 879 |

789, 798, 879, 897, 978, 987

3 Find the value.

(a) $500 + 80$ 580   (b) $508 - 304$ 204   (c) $800 - 40$ 760

(d) $544 + 381$ 925   (e) $623 - 218$ 405   (f) $487 + 255$ 742

(g) $527 - 248$ 279   (h) $804 - 317$ 487   (i) $700 - 281$ 419

4 Find the value.

(a) 7×5 35   (b) 6×2 12   (c) 5×4 20

(d) 2×9 18   (e) 8×10 80   (f) $45 \div 5$ 9

(g) $16 \div 2$ 8   (h) $20 \div 2$ 10   (i) $50 \div 10$ 5

182

182   Review 2

For the problems on page 183, students may use paper strips or sticky notes, or draw the bar models, if they need help figuring out the problem. They can also simply write an equation and solve. The bar models are provided as examples, not requirements.

5 — **14** are a mix of the four operations.

Ask students:

- What do we know?
- Do we have a total?
- Do we have parts?
- Do we have equal parts?
- Can we draw a picture?
- Can we act it out?
- What can we find first?
- Can we find an equation?

10 and **14** are two-step problems where step one is not given.

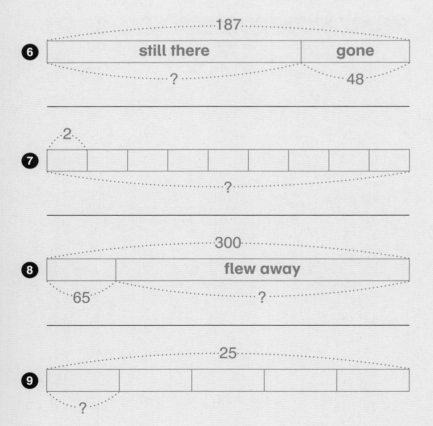

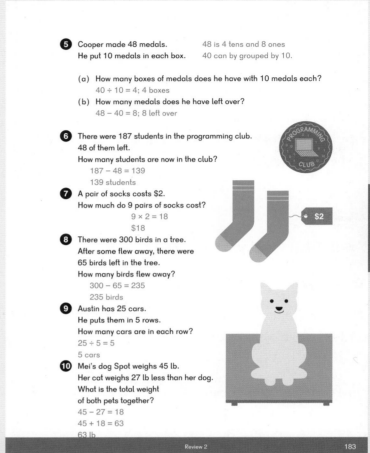

5 Cooper made 48 medals.   48 is 4 tens and 8 ones
He put 10 medals in each box.   40 can by grouped by 10.

(a) How many boxes of medals does he have with 10 medals each?
40 ÷ 10 = 4; 4 boxes

(b) How many medals does he have left over?
48 − 40 = 8; 8 left over

6 There were 187 students in the programming club.
48 of them left.
How many students are now in the club?
187 − 48 = 139
139 students

7 A pair of socks costs $2.
How much do 9 pairs of socks cost?
9 × 2 = 18
$18

8 There were 300 birds in a tree.
After some flew away, there were 65 birds left in the tree.
How many birds flew away?
300 − 65 = 235
235 birds

9 Austin has 25 cars.
He puts them in 5 rows.
How many cars are in each row?
25 ÷ 5 = 5
5 cars

10 Mei's dog Spot weighs 45 lb.
Her cat weighs 27 lb less than her dog.
What is the total weight of both pets together?
45 − 27 = 18
45 + 18 = 63
63 lb

183

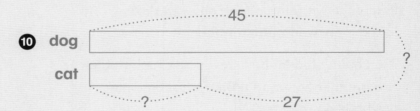

Teacher's Guide 2A Chapter 7   © 2017 Singapore Math Inc.

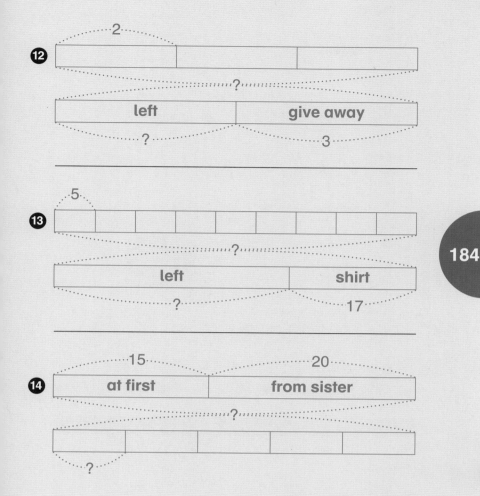

12

2

| left | give away |

? | 3

5

13

?

| left | shirt |

? | 17

---

15 | 20

14

| at first | from sister |

?

?

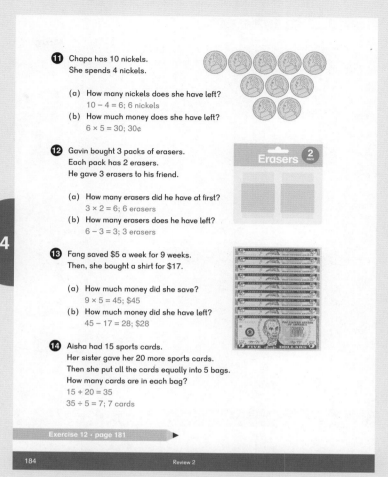

11 Chapa has 10 nickels.
She spends 4 nickels.

  (a) How many nickels does she have left?
     $10 - 4 = 6$; 6 nickels
  (b) How much money does she have left?
     $6 \times 5 = 30$; 30¢

12 Gavin bought 3 packs of erasers.
Each pack has 2 erasers.
He gave 3 erasers to his friend.

  (a) How many erasers did he have at first?
     $3 \times 2 = 6$; 6 erasers
  (b) How many erasers does he have left?
     $6 - 3 = 3$; 3 erasers

13 Fang saved $5 a week for 9 weeks.
Then, she bought a shirt for $17.

  (a) How much money did she save?
     $9 \times 5 = 45$; $45
  (b) How much money did she have left?
     $45 - 17 = 28$; $28

14 Aisha had 15 sports cards.
Her sister gave her 20 more sports cards.
Then she put all the cards equally into 5 bags.
How many cards are in each bag?
$15 + 20 = 35$
$35 \div 5 = 7$; 7 cards

> **Exercise 12 • page 181** ➤

184   Review 2

184

> **Exercise 12 • page 181** ➤

Chapter 7 Multiplication and Division of 2, 5, and 10

Exercise 1

Basics

1 Count by fives and complete the multiplication equations.

⚄	$1 \times 5 = $ 5
⚄ ⚄	$2 \times 5 = $ 10
⚄ ⚄ ⚄	$3 \times 5 = $ 15
⚄ ⚄ ⚄ ⚄	$4 \times 5 = $ 20
⚄ ⚄ ⚄ ⚄ ⚄	$5 \times 5 = $ 25
⚄ ⚄ ⚄ ⚄ ⚄ ⚄	$6 \times 5 = $ 30
⚄ ⚄ ⚄ ⚄ ⚄ ⚄ ⚄	$7 \times 5 = $ 35
⚄ ⚄ ⚄ ⚄ ⚄ ⚄ ⚄ ⚄	$8 \times 5 = $ 40
⚄ ⚄ ⚄ ⚄ ⚄ ⚄ ⚄ ⚄ ⚄	$9 \times 5 = $ 45
⚄ ⚄ ⚄ ⚄ ⚄ ⚄ ⚄ ⚄ ⚄ ⚄	$10 \times 5 = $ 50

2 The ones digit in all the products of 5 is either __5__ or __0__.

Practice

3 (a) $2 \times 5 = 5 + 5 = $ 10

(b) 3×5 is __5__ more than 2×5.

$3 \times 5 = $ 15

(c) 4×5 is 5 more than __3__ $\times 5$.

$4 \times 5 = $ 20

(d) $5 \times 5 = $ 20 $+ 5 = $ 25

(e) $10 \times 5 = $ 50

(f) $9 \times 5 = 50 - $ 5 $ = $ 45

(g) $8 \times 5 = $ 45 $ - 5 = $ 40

4 Each cake has 5 candles.

(a) Find how many are on 6 cakes.

6 $\times$ 5 $ = $ 30

(b) Find how many are on 7 cakes.

7 $\times$ 5 $ = $ 35

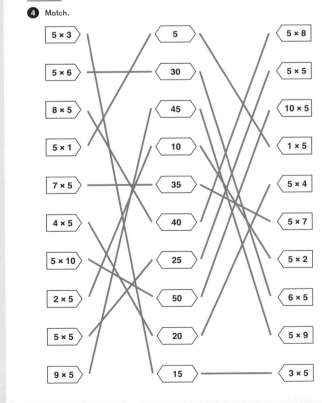

5 Circle products of 5.

(20)   14   (45)   12   (25)   (30)   28   (55)

Exercise 2

Basics

1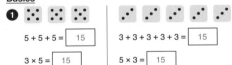

$5 + 5 + 5 = $ 15 $3 + 3 + 3 + 3 + 3 = $ 15

$3 \times 5 = $ 15 $5 \times 3 = $ 15

2 $7 \times 5 = $ 35

$5 \times 7 = $ 35

3

$6 \times 5 = $ 30		$5 \times 6 = $ 30
$2 \times 5 = $ 10		$5 \times 2 = $ 10
$10 \times 5 = $ 50		$5 \times 10 = $ 50
$1 \times 5 = $ 5		$5 \times 1 = $ 5
$9 \times 5 = $ 45		$5 \times 9 = $ 45
$4 \times 5 = $ 20		$5 \times 4 = $ 20
$8 \times 5 = $ 40		$5 \times 8 = $ 40
$3 \times 5 = $ 15		$5 \times 3 = $ 15
$7 \times 5 = $ 35		$5 \times 7 = $ 35
$5 \times 5 = $ 25		$5 \times 5 = $ 25

Practice

4 Match.

5×3 — 5 — 5×8
5×6 — 30 — 5×5
8×5 — 45 — 10×5
5×1 — 10 — 1×5
7×5 — 35 — 5×4
4×5 — 40 — 5×7
5×10 — 25 — 5×2
2×5 — 50 — 6×5
5×5 — 20 — 5×9
9×5 — 15 — 3×5

Exercise 3

Check

1 Count by fives and draw a path passing through the numbers in the correct order.

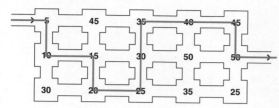

5	45	35	40	45
10	15	30	50	50
30	20	25	35	25

2 (a) 7 × 5 is ___5___ more than 6 × 5.

(b) 8 × 5 is 5 more than ___7___ × 5.

(c) 8 × 5 is 5 less than ___9___ × 5.

(d) 7 × 5 is ___10___ more than 5 × 5.

(e) 8 × 5 is 10 less than ___10___ × 5.

3 Write two different multiplication equations for the array.

$$5 \times 9 = 45$$
$$9 \times 5 = 45$$

4 (a) 5 × 6 = 30 (b) 5 × 2 = 10

(c) 9 × 5 = 45 (d) 1 × 5 = 5

(e) 5 × 3 = 15 (f) 7 × 5 = 35

(g) 2 × 5 = 10 (h) 4 × 5 = 20

(i) 5 × 7 = 35 (j) 5 × 10 = 50

(k) 10 × 5 = 50 (l) 5 × 9 = 45

(m) 5 × 5 = 25 (n) 8 × 5 = 40

(o) 5 × 4 = 20 (p) 3 × 5 = 15

(q) 5 × 8 = 40 (r) 6 × 5 = 30

5 (a) 5 × 5 = 25 (b) 5 × 3 = 15

(c) 8 × 5 = 40 (d) 5 × 6 = 30

6 There are 5 pretzels in each box.
How many pretzels are in 10 boxes?

10 × 5 = 50

There are ___50___ pretzels altogether.

7 There are 6 eggs in each carton.
How many eggs are in 5 cartons?

5 × 6 = 30

There are ___30___ eggs in 5 cartons.

8 Matthew saved $5 a week.
How much did he save in 7 weeks?

7 × 5 = 35

He saved $___35___ in 7 weeks.

9 Cora saved $8 a week.
How much did she save in 5 weeks?

5 × 8 = 40

She saved $___40___ in 5 weeks.

10 Andrei's puppy weighed 5 lb.
It is now full grown and is 9 times heavier.
How much does Andrei's dog now weigh?

9 × 5 = 45

His dog now weighs ___45___ pounds.

Challenge

11 Rowan has 9 stickers on each of 5 pages.
Taylor has 5 stickers on each of 7 pages.
Who has more stickers and how much more?
Rowan: 5 × 9 = 45
Taylor: 7 × 5 = 35
45 − 35 = 10

___Rowan___ has ___10___ more stickers than ___Taylor___.

12 5 × 5 = ★ + ★ + 5

The ★ stands for ___10___.

13 Ximena has 10 pads of watercolor paper.
Each pad has 5 sheets of watercolor paper.
She has used all the sheets from 4 pads and 3 sheets from the 5th pad.

(a) How many sheets has she used?
4 × 5 = 20
20 + 3 = 23

She has used ___23___ sheets.

(b) How many sheets does she have left?
10 × 5 = 50 (total sheets)   Or, 5 pads and 2 sheets left.
50 − 23 = 27   5 × 5 + 2 = 27

She has ___27___ sheets left.

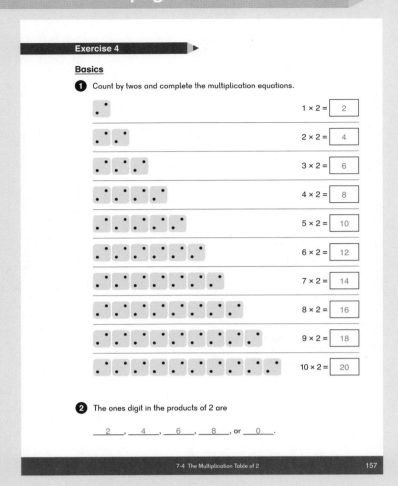

Exercise 4

Basics

1 Count by twos and complete the multiplication equations.

$1 \times 2 =$ 2

$2 \times 2 =$ 4

$3 \times 2 =$ 6

$4 \times 2 =$ 8

$5 \times 2 =$ 10

$6 \times 2 =$ 12

$7 \times 2 =$ 14

$8 \times 2 =$ 16

$9 \times 2 =$ 18

$10 \times 2 =$ 20

2 The ones digit in the products of 2 are

__2__, __4__, __6__, __8__, or __0__.

7-4 The Multiplication Table of 2   157

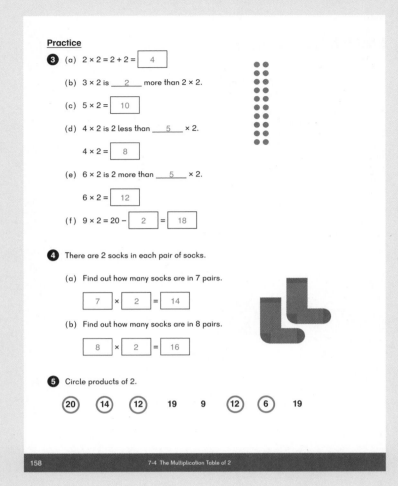

Practice

3 (a) $2 \times 2 = 2 + 2 =$ 4

(b) 3×2 is __2__ more than 2×2.

(c) $5 \times 2 =$ 10

(d) 4×2 is 2 less than __5__ $\times 2$.

$4 \times 2 =$ 8

(e) 6×2 is 2 more than __5__ $\times 2$.

$6 \times 2 =$ 12

(f) $9 \times 2 = 20 -$ 2 $=$ 18

4 There are 2 socks in each pair of socks.

(a) Find out how many socks are in 7 pairs.

7 $\times$ 2 $=$ 14

(b) Find out how many socks are in 8 pairs.

8 $\times$ 2 $=$ 16

5 Circle products of 2.

(20) (14) (12) 19 9 (12) (6) 19

158   7-4 The Multiplication Table of 2

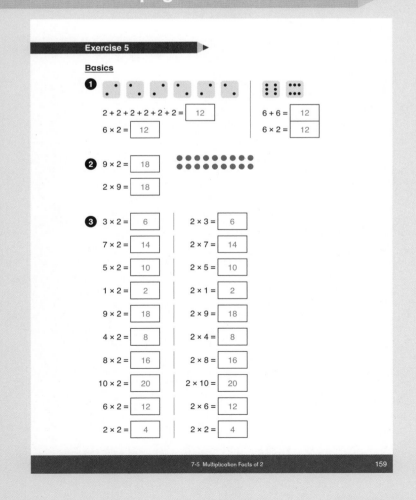

Exercise 5

Basics

1

$2 + 2 + 2 + 2 + 2 + 2 =$ 12

$6 \times 2 =$ 12

$6 + 6 =$ 12

$6 \times 2 =$ 12

2 $9 \times 2 =$ 18

$2 \times 9 =$ 18

3 $3 \times 2 =$ 6 | $2 \times 3 =$ 6

$7 \times 2 =$ 14 | $2 \times 7 =$ 14

$5 \times 2 =$ 10 | $2 \times 5 =$ 10

$1 \times 2 =$ 2 | $2 \times 1 =$ 2

$9 \times 2 =$ 18 | $2 \times 9 =$ 18

$4 \times 2 =$ 8 | $2 \times 4 =$ 8

$8 \times 2 =$ 16 | $2 \times 8 =$ 16

$10 \times 2 =$ 20 | $2 \times 10 =$ 20

$6 \times 2 =$ 12 | $2 \times 6 =$ 12

$2 \times 2 =$ 4 | $2 \times 2 =$ 4

7-5 Multiplication Facts of 2   159

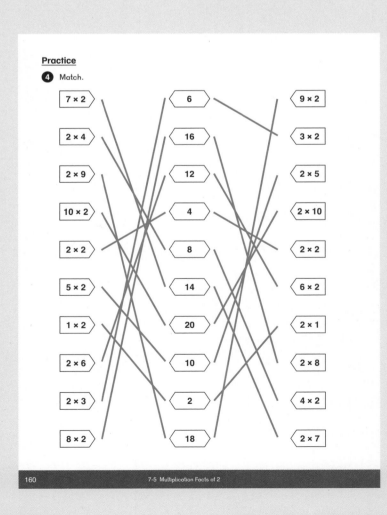

Practice

4 Match.

160   7-5 Multiplication Facts of 2

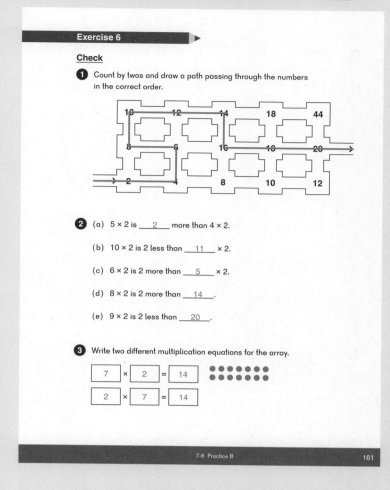

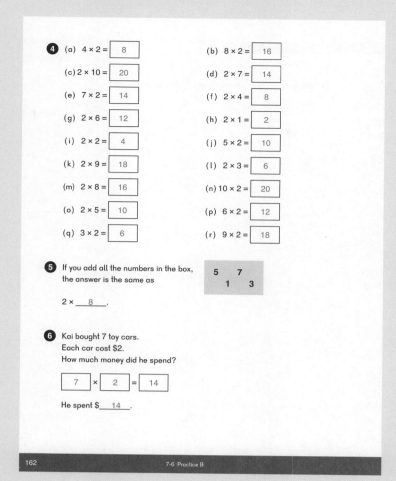

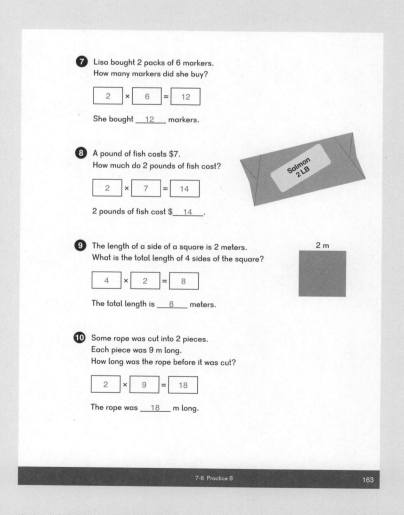

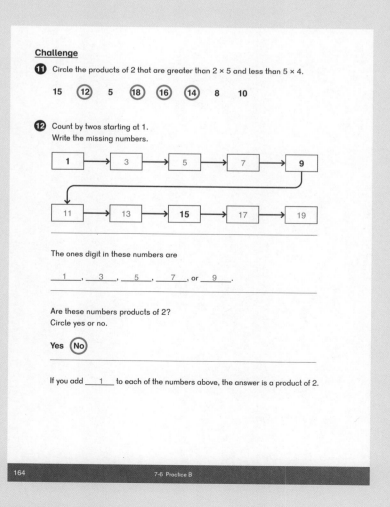

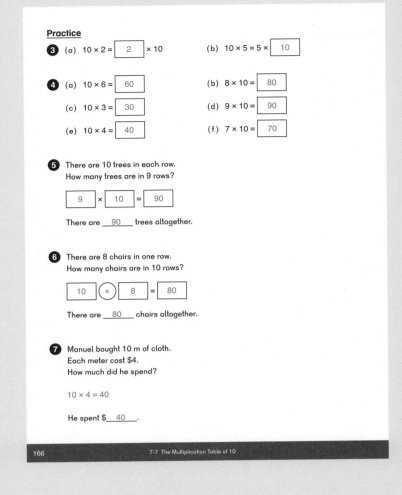

Exercise 7

Basics

1 Count by tens and complete the multiplication equations.

🪙	1 × 10 = **10**
🪙🪙	2 × 10 = **20**
🪙🪙🪙	3 × 10 = **30**
🪙🪙🪙🪙	4 × 10 = **40**
🪙🪙🪙🪙🪙	5 × 10 = **50**
🪙🪙🪙🪙🪙🪙	6 × 10 = **60**
🪙🪙🪙🪙🪙🪙🪙	7 × 10 = **70**
🪙🪙🪙🪙🪙🪙🪙🪙	8 × 10 = **80**
🪙🪙🪙🪙🪙🪙🪙🪙🪙	9 × 10 = **90**
🪙🪙🪙🪙🪙🪙🪙🪙🪙🪙	10 × 10 = **100**

2 The ones digit in all the products of 10 is **0**.

Practice

3 (a) 10 × 2 = **2** × 10   (b) 10 × 5 = 5 × **10**

4 (a) 10 × 6 = **60**   (b) 8 × 10 = **80**

(c) 10 × 3 = **30**   (d) 9 × 10 = **90**

(e) 10 × 4 = **40**   (f) 7 × 10 = **70**

5 There are 10 trees in each row.
How many trees are in 9 rows?

9 × **10** = **90**

There are **90** trees altogether.

6 There are 8 chairs in one row.
How many chairs are in 10 rows?

10 ⊗ **8** = **80**

There are **80** chairs altogether.

7 Manuel bought 10 m of cloth.
Each meter cost $4.
How much did he spend?

10 × 4 = 40

He spent $**40**.

Exercise 8

Basics

1 (a) Mei has 14 socks to sort into pairs.
How many pairs of socks does she have?

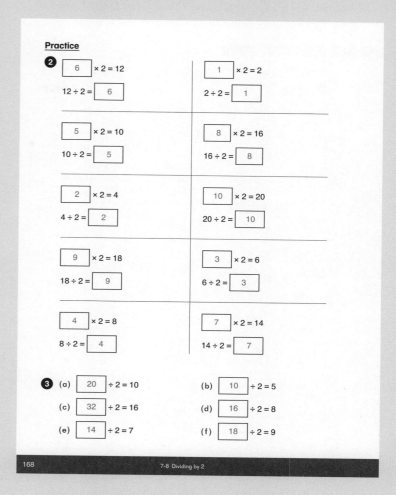

___7___ groups of 2 is 14.

☐7☐ × 2 = 14

14 grouped by 2 is ___7___ groups.

14 ÷ 2 = ☐7☐

She has ___7___ pairs of socks.

(b) Sofia links up 14 train cars into two trains with the same number of cars.
How many train cars are in each train?

2 groups of ___7___ is 14.

2 × ☐7☐ = 14

14 shared between 2 groups is ___7___ in each group.

14 ÷ 2 = ☐7☐

There are ___7___ train cars in each train.

Practice

2

☐6☐ × 2 = 12	☐1☐ × 2 = 2
12 ÷ 2 = ☐6☐	2 ÷ 2 = ☐1☐
☐5☐ × 2 = 10	☐8☐ × 2 = 16
10 ÷ 2 = ☐5☐	16 ÷ 2 = ☐8☐
☐2☐ × 2 = 4	☐10☐ × 2 = 20
4 ÷ 2 = ☐2☐	20 ÷ 2 = ☐10☐
☐9☐ × 2 = 18	☐3☐ × 2 = 6
18 ÷ 2 = ☐9☐	6 ÷ 2 = ☐3☐
☐4☐ × 2 = 8	☐7☐ × 2 = 14
8 ÷ 2 = ☐4☐	14 ÷ 2 = ☐7☐

3 (a) ☐20☐ ÷ 2 = 10   (b) ☐10☐ ÷ 2 = 5

(c) ☐32☐ ÷ 2 = 16   (d) ☐16☐ ÷ 2 = 8

(e) ☐14☐ ÷ 2 = 7   (f) ☐18☐ ÷ 2 = 9

4 Match.

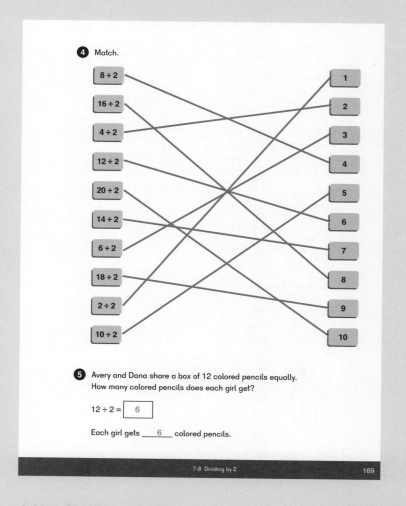

8 ÷ 2	1
16 ÷ 2	2
4 ÷ 2	3
12 ÷ 2	4
20 ÷ 2	5
14 ÷ 2	6
6 ÷ 2	7
18 ÷ 2	8
2 ÷ 2	9
10 ÷ 2	10

5 Avery and Dana share a box of 12 colored pencils equally.
How many colored pencils does each girl get?

12 ÷ 2 = ☐6☐

Each girl gets ___6___ colored pencils.

6 Mrs. Huber has 20 roses.
She puts 2 roses in each vase.
How many vases does she use?

20 ÷ 2 = ☐10☐

She uses ___10___ vases.

7 Divide 16 children equally into 2 groups.
How many children are in each group?

☐16☐ ⊘ ☐2☐ = ☐8☐

There are ___8___ children in each group.

8 Mr. Jung has a rope that is 18 m long.
He cuts it into equal pieces that are each 2 m long.
How many pieces of rope does he have?

18 ÷ 2 = 9

He has ___9___ pieces of rope.

Challenge

9 A tailor bought 14 m of cloth on Tuesday.
Starting on Wednesday, he cut off 2 m of cloth each day to use.
On Sunday, he did not cut any cloth.
On what day of the week did he make the last cut?

He only has to make 6 cuts to cut it into 7 pieces.
6th day counting Wednesday but skipping Sunday, is Tuesday.
Students can draw a picture.

He made the last cut on ___Tuesday___.

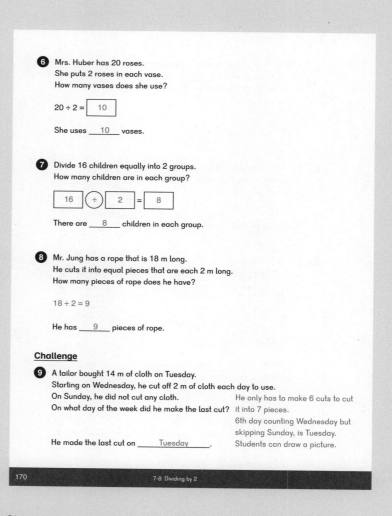

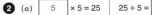

Exercise 9

Basics

1 Alex has 40 marbles.

(a) If he puts 5 marbles in each bag, how many bags will he have?

$\boxed{8} \times 5 = 40$ | $40 \div 5 = \boxed{8}$

He will have __8__ bags.

(b) If he puts them equally in 5 bags, how many will be in each bag?

$5 \times \boxed{8} = 40$ | $40 \div 5 = \boxed{8}$

Each bag will have __8__ marbles.

Practice

2 (a) $\boxed{5} \times 5 = 25$ | $25 \div 5 = \boxed{5}$

(b) $\boxed{2} \times 5 = 10$ | $10 \div 5 = \boxed{2}$

(c) $\boxed{9} \times 5 = 45$ | $45 \div 5 = \boxed{9}$

(d) $\boxed{4} \times 5 = 20$ | $20 \div 5 = \boxed{4}$

(e) $\boxed{7} \times 5 = 35$ | $35 \div 5 = \boxed{7}$

(f) $\boxed{3} \times 5 = 15$ | $15 \div 5 = \boxed{3}$

(g) $\boxed{6} \times 5 = 30$ | $30 \div 5 = \boxed{6}$

(h) $\boxed{1} \times 5 = 5$ | $5 \div 5 = \boxed{1}$

7-9 Dividing by 5 and 10   171

3 (a) $\boxed{4} \times 10 = 40$ $40 \div 10 = \boxed{4}$

(b) $\boxed{9} \times 10 = 90$ $90 \div 10 = \boxed{9}$

(c) $\boxed{7} \times 10 = 70$ $70 \div 10 = \boxed{7}$

4 Match.

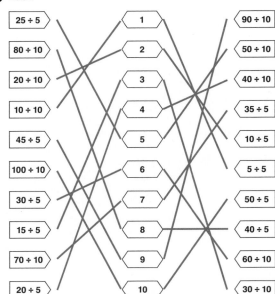

172   7-9 Dividing by 5 and 10

5 80 children line up in 10 rows for a performance.
There are the same number of children in each row.
How many children are in each row?

$\boxed{80} \div \boxed{10} = \boxed{8}$

__8__ children are in each row.

6 10 children share 90 crayons equally.
How many crayons does each child get?

$\boxed{90} \div \boxed{10} = \boxed{9}$

Each child gets __9__ crayons.

7 5 shirts cost $25.
Each shirt costs the same amount.

(a) How much does each shirt cost?

$\boxed{25} \div \boxed{5} = \boxed{5}$

Each shirt costs $__5__.

(b) Grant spent $30 on these shirts.
How many shirts did he buy?

$\boxed{30} \div \boxed{5} = \boxed{6}$

He bought __6__ shirts.

7-9 Dividing by 5 and 10   173

8 (a) $\boxed{40} \div 10 = 4$ (b) $\boxed{20} \div 5 = 4$

(c) $\boxed{80} \div 10 = 8$ (d) $\boxed{40} \div 5 = 8$

(e) $\boxed{70} \div 10 = 7$ (f) $\boxed{35} \div 5 = 7$

Challenge

9 Mr. Phillips has a rope that is 27 m long.
How many 5-m pieces can he cut from the rope?   5 pieces
How long will the leftover piece be?   2 m
$27 \div 5$ is 5 with 2 left over.
Check: $5 \times 5 = 25$
$25 + 2 = 27$

He can cut __5__ pieces of rope that are each 5 m long.

The leftover piece is __2__ m long.

10 Complete the number puzzle.

10	×	4	=	40
÷		÷		÷
5	×	2	=	10
=		=		=
2	×	2	=	4

174   7-9 Dividing by 5 and 10

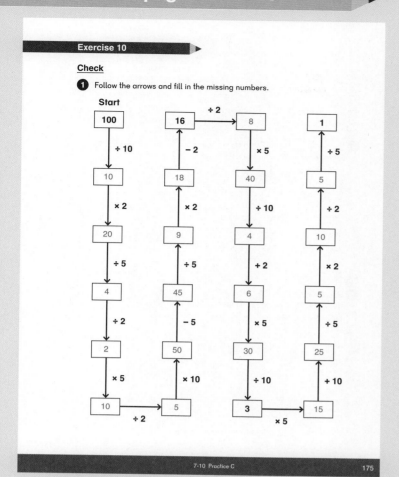

Exercise 10

Check

1 Follow the arrows and fill in the missing numbers.

Start

| 100 | 16 | ÷2 → 8 | 1 |

(Column 1) 100 → ÷10 → 10 → ×2 → 20 → ÷5 → 4 → ÷2 → 2 → ×5 → 10 → ÷2 → 5

(Column 2) 16 ← −2 ← 18 ← ×2 ← 9 ← ÷5 ← 45 ← −5 ← 50 ← ×10 ← 5

(Column 3) 8 → ×5 → 40 → ÷10 → 4 → ÷2 → 6 → ×5 → 30 → ÷10 → 3 → ×5 → 15

(Column 4) 1 ← ÷5 ← 5 ← ÷2 ← 10 ← ×2 ← 5 ← ÷5 ← 25 ← +10 ← 15

7-10 Practice C   175

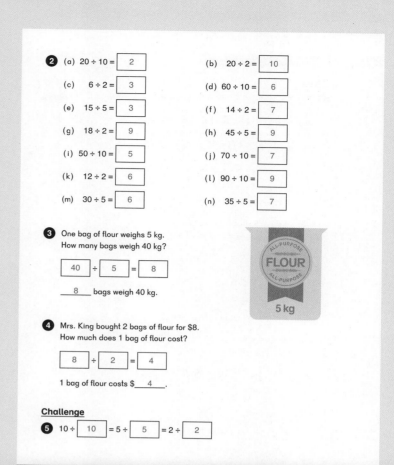

2 (a) 20 ÷ 10 = [2]   (b) 20 ÷ 2 = [10]

(c) 6 ÷ 2 = [3]   (d) 60 ÷ 10 = [6]

(e) 15 ÷ 5 = [3]   (f) 14 ÷ 2 = [7]

(g) 18 ÷ 2 = [9]   (h) 45 ÷ 5 = [9]

(i) 50 ÷ 10 = [5]   (j) 70 ÷ 10 = [7]

(k) 12 ÷ 2 = [6]   (l) 90 ÷ 10 = [9]

(m) 30 ÷ 5 = [6]   (n) 35 ÷ 5 = [7]

3 One bag of flour weighs 5 kg.
How many bags weigh 40 kg?

[40] ÷ [5] = [8]

___8___ bags weigh 40 kg.

FLOUR
ALL-PURPOSE
5 kg

4 Mrs. King bought 2 bags of flour for $8.
How much does 1 bag of flour cost?

[8] ÷ [2] = [4]

1 bag of flour costs $___4___.

Challenge

5 10 ÷ [10] = 5 ÷ [5] = 2 ÷ [2]

176   7-10 Practice C

Exercise 11

Basics

1 Phyllis mixes together 26 kg of almonds and 24 kg of cashews.

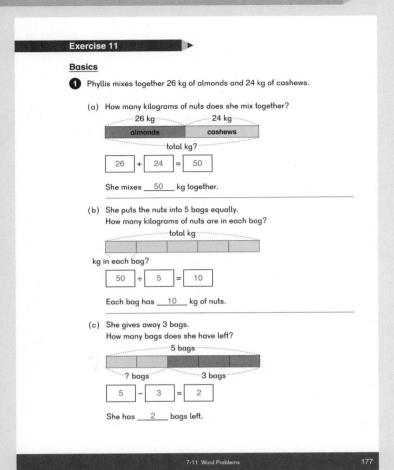

(a) How many kilograms of nuts does she mix together?

26 kg   24 kg
[almonds] [cashews]
total kg?

$26 + 24 = 50$

She mixes __50__ kg together.

(b) She puts the nuts into 5 bags equally.
How many kilograms of nuts are in each bag?

total kg
kg in each bag?

$50 ÷ 5 = 10$

Each bag has __10__ kg of nuts.

(c) She gives away 3 bags.
How many bags does she have left?

5 bags
? bags   3 bags

$5 - 3 = 2$

She has __2__ bags left.

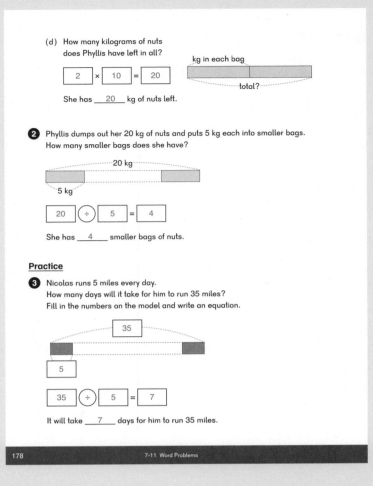

(d) How many kilograms of nuts
does Phyllis have left in all?

kg in each bag

$2 × 10 = 20$

total?

She has __20__ kg of nuts left.

2 Phyllis dumps out her 20 kg of nuts and puts 5 kg each into smaller bags.
How many smaller bags does she have?

20 kg
5 kg

$20 ÷ 5 = 4$

She has __4__ smaller bags of nuts.

Practice

3 Nicolas runs 5 miles every day.
How many days will it take for him to run 35 miles?
Fill in the numbers on the model and write an equation.

35
5

$35 ÷ 5 = 7$

It will take __7__ days for him to run 35 miles.

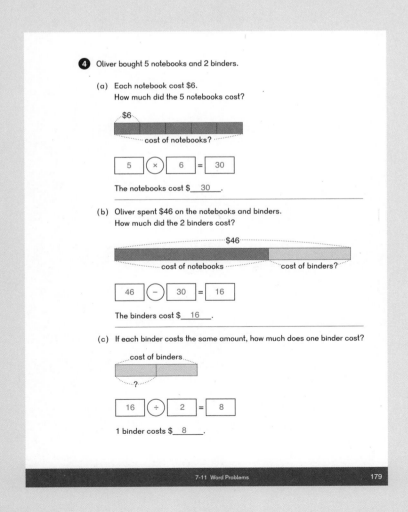

4 Oliver bought 5 notebooks and 2 binders.

(a) Each notebook cost $6.
How much did the 5 notebooks cost?

$6
cost of notebooks?

$5 × 6 = 30$

The notebooks cost $ __30__ .

(b) Oliver spent $46 on the notebooks and binders.
How much did the 2 binders cost?

$46
cost of notebooks   cost of binders?

$46 - 30 = 16$

The binders cost $ __16__ .

(c) If each binder costs the same amount, how much does one binder cost?

cost of binders
?

$16 ÷ 2 = 8$

1 binder costs $ __8__ .

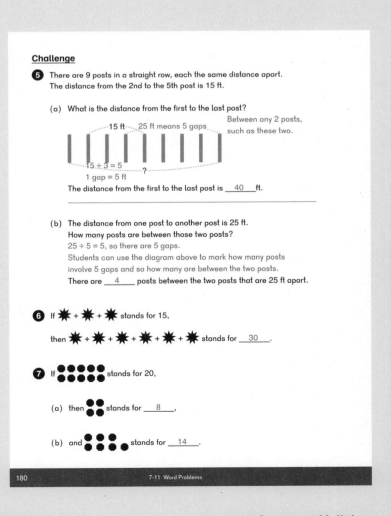

Challenge

5 There are 9 posts in a straight row, each the same distance apart.
The distance from the 2nd to the 5th post is 15 ft.

(a) What is the distance from the first to the last post?

15 ft   25 ft means 5 gaps
Between any 2 posts, such as these two.

$15 ÷ 3 = 5$?
1 gap = 5 ft

The distance from the first to the last post is __40__ ft.

(b) The distance from one post to another post is 25 ft.
How many posts are between those two posts?
$25 ÷ 5 = 5$, so there are 5 gaps.
Students can use the diagram above to mark how many posts
involve 5 gaps and so how many are between the two posts.
There are __4__ posts between the two posts that are 25 ft apart.

6 If ✹ + ✹ + ✹ stands for 15,

then ✹ + ✹ + ✹ + ✹ + ✹ + ✹ stands for __30__ .

7 If ●●●●● / ●●●●● stands for 20,

(a) then ●● / ●● stands for __8__ ,

(b) and ●●● / ●●● ● stands for __14__ .

Exercise 12 • pages 181–184

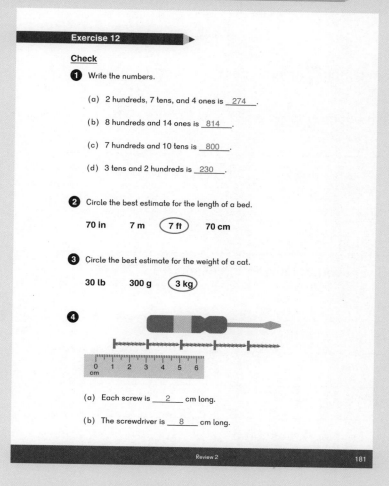

Exercise 12

Check

1 Write the numbers.

    (a) 2 hundreds, 7 tens, and 4 ones is __274__.

    (b) 8 hundreds and 14 ones is __814__.

    (c) 7 hundreds and 10 tens is __800__.

    (d) 3 tens and 2 hundreds is __230__.

2 Circle the best estimate for the length of a bed.

    70 in   7 m   (7 ft)   70 cm

3 Circle the best estimate for the weight of a cat.

    30 lb   300 g   (3 kg)

4

    (a) Each screw is __2__ cm long.

    (b) The screwdriver is __8__ cm long.

Review 2 — page 181

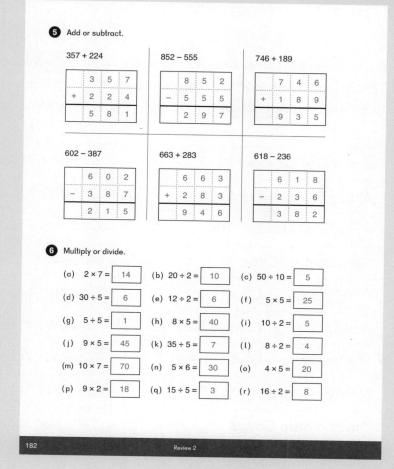

5 Add or subtract.

357 + 224	852 − 555	746 + 189
3 5 7	8 5 2	7 4 6
+ 2 2 4	− 5 5 5	+ 1 8 9
5 8 1	2 9 7	9 3 5

602 − 387	663 + 283	618 − 236
6 0 2	6 6 3	6 1 8
− 3 8 7	+ 2 8 3	− 2 3 6
2 1 5	9 4 6	3 8 2

6 Multiply or divide.

    (a) $2 \times 7 = $ __14__   (b) $20 \div 2 = $ __10__   (c) $50 \div 10 = $ __5__

    (d) $30 \div 5 = $ __6__   (e) $12 \div 2 = $ __6__   (f) $5 \times 5 = $ __25__

    (g) $5 \div 5 = $ __1__   (h) $8 \times 5 = $ __40__   (i) $10 \div 2 = $ __5__

    (j) $9 \times 5 = $ __45__   (k) $35 \div 5 = $ __7__   (l) $8 \div 2 = $ __4__

    (m) $10 \times 7 = $ __70__   (n) $5 \times 6 = $ __30__   (o) $4 \times 5 = $ __20__

    (p) $9 \times 2 = $ __18__   (q) $15 \div 5 = $ __3__   (r) $16 \div 2 = $ __8__

Review 2 — page 182

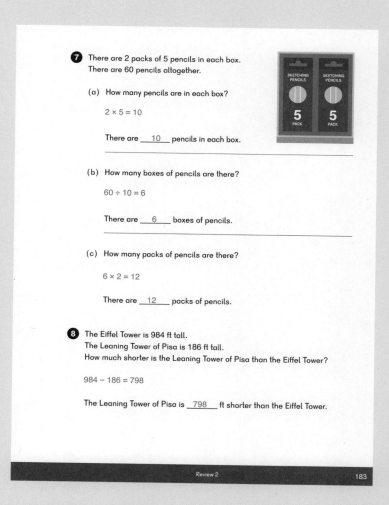

7 There are 2 packs of 5 pencils in each box. There are 60 pencils altogether.

    (a) How many pencils are in each box?

        $2 \times 5 = 10$

        There are __10__ pencils in each box.

    (b) How many boxes of pencils are there?

        $60 \div 10 = 6$

        There are __6__ boxes of pencils.

    (c) How many packs of pencils are there?

        $6 \times 2 = 12$

        There are __12__ packs of pencils.

8 The Eiffel Tower is 984 ft tall.
The Leaning Tower of Pisa is 186 ft tall.
How much shorter is the Leaning Tower of Pisa than the Eiffel Tower?

    $984 − 186 = 798$

    The Leaning Tower of Pisa is __798__ ft shorter than the Eiffel Tower.

Review 2 — page 183

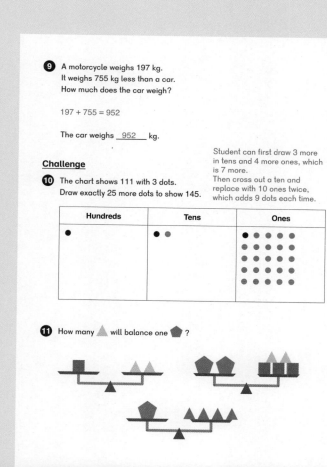

9 A motorcycle weighs 197 kg.
It weighs 755 kg less than a car.
How much does the car weigh?

    $197 + 755 = 952$

    The car weighs __952__ kg.

Challenge

10 The chart shows 111 with 3 dots.
Draw exactly 25 more dots to show 145.

Student can first draw 3 more in tens and 4 more ones, which is 7 more.
Then cross out a ten and replace with 10 ones twice, which adds 9 dots each time.

Hundreds	Tens	Ones
●	● ●	●●●●● ●●●●● ●●●●● ●●●●● ●●●●●

11 How many 🔺 will balance one ⬠ ?

Review 2 — page 184

Teacher's Guide 2A Chapter 7

Blackline Masters for 2A

All Blackline Masters used in the guide can be downloaded from dimensionsmath.com.
This lists BLM used in the **Think** and **Learn** sections.
BLMs used in **Activities** are listed in the Activity Materials within each chapter.

Array Dot Cards - 2s	**Chapter 7:** Lesson 4
Array Dot Cards - 5s	**Chapter 7:** Lesson 1
Array Dot Cards - 10s	**Chapter 7:** Lesson 7
Centimeter Ruler	**Chapter 4:** Lesson 1, Lesson 2, Lesson 7, Lesson 8
Multiplication Chart - 2s	**Chapter 7:** Lesson 4
Multiplication Chart - 5s	**Chapter 7:** Lesson 1
Multiplication Chart - 10s	**Chapter 7:** Lesson 7
Number Cards	**Chapter 1:** Lesson 3
Place-value Cards	**Chapter 1:** Lesson 1, Lesson 4, Lesson 5